T0132584

Hidden
Natural
Histories

HERBS

Kim Hurst is the author of *Herbs and the Kitchen Garden*.

First published in the United States of America in 2015 by
The University of Chicago Press Chicago 60637

Every effort has been made to contact the copyright holders of material in this book. However, where an omission has occurred, the publisher will gladly include acknowledgment in any future edition.

24 23 22 21 20 19 18 17 16 15   1 2 3 4 5

ISBN-13: 978-0-226-27117-0   (paper)
ISBN-13: 978-0-226-24699-4   (e-book)
DOI: 10.7208/chicago/9780226246994.0001

Library of Congress Cataloging-in-Publication Data

Hurst, Kim, 1956– author.
  Hidden natural histories. Herbs / Kim Hurst.
    pages cm
  ISBN 978-0-226-27117-0 (paperback : alkaline paper) — ISBN 0-226-27117-X (paperback : alkaline paper) — ISBN 978-0-226-24699-4 (e-book) — ISBN 0-226-24699-X (e-book) 1. Herbs—Utilization. I. Title.
  GT5164.H86 2015
  635'.7—dc23

                                                                      2014037016

This book was designed and produced by
Quintessence Editions Ltd. The Old Brewery, 6 Blundell Street, London N7 9BH

Project Editor	Zoë Smith
Designer	Isabel Eeles
Editors	Frank Ritter, Fiona Plowman
Production Manager	Anna Pauletti
Editorial Director	Jane Laing
Publisher	Mark Fletcher

Color reproduction by Colourscan Print Co Pte Ltd, Singapore
Printed and bound in China by Shanghai Offset Printing Products Ltd.

⊗ This paper meets the requirements of ANSI/NISO Z39.48-1992 (Permanence of Paper)

The publisher and author are not responsible for any adverse effects or consequences resulting from the use of suggestions, preparations, or procedures discussed in this book. Should the reader have any questions concerning the appropriateness of any procedures or preparation mentioned, the author and publisher strongly suggest consulting a professional healthcare advisor.

Hidden Natural Histories

KIM HURST

HERBS

THE SECRET PROPERTIES OF **150 PLANTS**

THE UNIVERSITY OF CHICAGO PRESS

CHICAGO

Herbs by Common Name

Contents

How to Use This Book

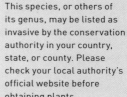

Latin name of herb, organized alphabetically

Symbols depicting herb parts that are used

Cultivation details

Common name(s) of herb

Introduction

Descriptive panels

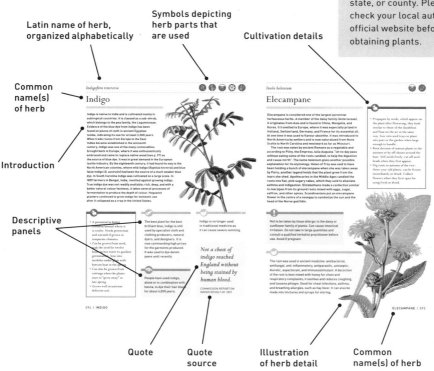

Indigofera tinctoria

Indigo

Indigo is native to India and is cultivated mainly in subtropical countries. It is classed as a sub-shrub, which belongs to the pea family, the Leguminosae. Evidence of the blue dye from indigo has been found on pieces of cloth in ancient Egyptian tombs, indicating its use for at least 3,000 years. When trade routes from Europe to the East Indies became established in the sixteenth century, indigo was one of the many commodities brought back to Europe, where it was enthusiastically embraced and came to replace native woad (see p. 97) as the source of blue dye. It was in great demand in the European textile industry. By the eighteenth century, it had found its way to the North American colonies, where wild indigo (*Baptisia tinctoria*) and blue false indigo (*B. australis*) had been the source of a much weaker blue dye. In South Carolina indigo was cultivated on a large scale. In 1859 farmers in Bengal, India, revolted against growing indigo. True indigo dye was not readily available; rich, deep, and with a better natural colour fastness, it takes several processes of fermentation to produce the depth of colour. Huguenot planters continued to grow indigo for domestic use after it collapsed as a crop in the United States.

• A perennial in hotter climates/annual where it is cooler. Needs protection and warmth if grown in temperate climates.
• Can be grown from seed; soak the seed for twelve hours in hot water to quicken germination. Sow into modules under glass with bottom heat in the spring.
• Can also be grown from cuttings when the plants start to "grow away" in late spring.
• Grows well in nutrient deficient soil.

The best plant for the best brilliant blue, indigo is still used by specialist cloth and clothing producers, natural dyers, and designers. It is now commanding high prices for the garments produced. It was used to dye denim jeans until recently

People have used indigo, alone or in combination with henna, to dye their hair black for about 4,000 years.

Indigo is no longer used in traditional medicine as it can cause severe vomiting.

Not a chest of indigo reached England without being stained by human blood.

COMMISSION REPORT ON INDIGO REVOLT OF 1859

Inula helenium

Elecampane

Elecampane is considered one of the largest perennial herbaceous herbs. A member of the daisy family (Asteraceae), it originates from Asia and is found in China, Mongolia, and Korea. It travelled to Europe, where it was especially prized in Holland, Switzerland, Germany, and France for its essential oil. At one time it was used to flavour absinthe. It was introduced to North America by settlers and is now naturalized from Nova Scotia to North Carolina and westward as far as Missouri. The root was eaten by ancient Romans as a vegetable and according to Pliny, the Empress Julia Augusta "let no day pass without eating some of the roots candied, to help the digestion and cause mirth". The name helenium gives another possible explanation for its etymology. Helen of Troy was said to have been holding a bunch of elecampane when she was taken away by Paris; another legend holds that the plant grew from the tears she shed. Apothecaries in the Middle Ages candied the roots into flat, pink sugary cakes, which they sold to alleviate asthma and indigestion. Elizabethans made a confection similar to marzipan from its ground roots mixed with eggs, sugar, saffron, and other spices. Scandinavians put an elecampane flower in the centre of a nosegay to symbolize the sun and the head of the Norse god Odin.

• Propagate by seeds, which appear on the plant after flowering; they look similar to those of the dandelion and float on the air in the same way. Sow into seed trays or plant into pots or the garden when large enough to handle.
• Root division of mature plants in the autumn or by off-shoots around the base. Self-seeds freely; cut off seed heads when they first appear.
• Dig roots in autumn of the two-three-year-old plants; can be frozen immediately or dried. Collect flowers when they first open for using fresh or dried.

Not to be taken by those allergic to the daisy or sunflower family of plants. Can cause intestinal irritation. Do not take in large quantities and consult a qualified herbalist practitioner before use. Avoid if pregnant.

The root was used in ancient medicine: antibacterial, antifungal, anti-inflammatory, antiparasitic, antiseptic, diuretic, expectorant, and immunostimulant. A decoction of the root is best mixed with honey for chest and respiratory complaints; it soothes and reduces coughing, and loosens phlegm. Good for chest infections, asthma, and breathing allergies, such as hay fever. It can also be made into tinctures and syrups for storing.

Quote

Quote source

Illustration of herb detail

Common name(s) of herb

Herb parts used

 Leaves

 Stalks/stems

 Flowers

 Rhizomes

 Wood

 Roots

 Fruits

 Seeds

 Bulbs

 Bark

Key to descriptive panels

 Culinary

 Medicinal

 Domestic

 Folklore & religion

 Cosmetic

 Warning

 Companion plants

 Craft

Herbs are at the forefront of the current revolution in cooking and international cuisine, with a much wider range of exotic plants contributing their flavors to Western ingredients.

Whole branches of the modern pharmacopoeia are founded on ancient discoveries relating to the abilities of particular plants to relieve, and sometimes cure, medical disorders.

Over history, people have strewn herbs in their homes to sweeten the air and repel insects, and created life-enhancing products with them, such as scented sachets and furniture polish.

Magical herbs lie at the heart of folklore and feature in early beliefs about the supernatural world. Some herbal charms and amulets were believed to protect against witchcraft and evil.

Since ancient times, people have found cosmetic uses for herbs, using them in lotions, salves, and soaps. The plants add both fragrance and curative properties to thousands of preparations.

Just as some herbs are powerful healers, others are poisonous and dangerous. Indeed, some may be both, being beneficial in tiny doses but harmful in excess or when wrongly prepared.

Planted alongside food crops or ornamental plants, some herbs are valuable "companion plants," able to increase their neighbors' vitality, heal them, or protect them from pests or diseases.

Craftspeople have long used herbs as raw materials, weaving flax or nettle fibers into fabrics, for example, and coloring them with dyes extracted from herbs such as madder and woad.

Introduction *Kim Hurst*

Many thousands of years ago, our ancient ancestors began to learn that certain plants brought special benefits in terms of nutrition, natural remedies, or the desirable commodities that they yielded. But herbs came to be regarded as far more than simple commodities. Because many could effect seemingly miraculous recoveries from disease and injury, they were revered as forces for good, not only in the material world but also in the battle against the devil and evil. Herbalists and apothecaries attained the status of magicians, and people were as likely to ask for a potion to attract a lover, protect against wild animals, or banish nightmares as request a remedy for cramp or rheumatism.

Plants for humankind's ease

Many of our uses of herbs are hidden, in the sense that they are taken for granted, or not realized, hence the idea of the "hidden history." This book tells the stories of 150 of the most remarkable plants used through the ages for their properties: culinary, medicinal, cosmetic, domestic, craft, magical, or as natural companion plants. Often, the herbs most used for cooking and medicine were also the ones most revered for their supernatural powers. Rosemary, for example, was burned as incense to cleanse and purify a room, especially a sickroom. It was hung over the door to keep thieves from entering the house; worn to aid the memory and preserve youth; bound to the right arm to cure depression; and even grown to attract helpful elves to the garden. Yarrow, also known as soldier's wort, was said to have been carried into battle by warriors, not just as a self-medication but also to protect them from harm and make them more courageous. Many herbs had unexpected uses: garlic was used to rid gardens of moles; eyebright was thought to increase psychic powers; and fennel seeds were eaten in the Middle Ages to allay hunger and as an antidote to food poisoning. Common sage, once known as *Salvia salvatrix* or "sage the savior," was reputed to promote longevity in people who drank it in daily infusions. Meadowsweet, with its creamy white, frothy, scented flowers was sacred to the Druids, and Queen Elizabeth I of England favored it as an uplifting strewing herb.

A remarkable aspect of the hidden history of herbs is that ancient civilizations, developing across the world and in some cases totally isolated from one another, independently discovered the powers of the plants around them. For example, the indigenous people of Australia used a tea of eucalyptus leaves to treat pains, sinus congestion, fever, and colds, while Native Americans used ipecac (*Gillenia trifoliata*) as a laxative and emetic. European settlers would call the plant "Indian physic" in recognition of its widespread use in North America.

The desire to compile information about the properties of plants in an accessible form is ancient. The Roman philosopher and naturalist Pliny the Elder (23–79 CE) wrote *Naturalis Historia,* an encyclopaedia that included several volumes on medical botany. The Greek physician Pedanius Dioscorides (*c.* 40–90 CE) wrote *De Materia Medica,* a five-volume encyclopedia about herbal medicine and medicinal substances. Observations from these authors and others became interwoven in the minds of the common people with superstition and folklore.

The first century also saw the emergence of the highly influential Doctrine of Signatures, although it probably existed in an earlier form in ancient China. The doctrine suggested that easily visible aspects of plants, such as the shape or color of their leaves, flowers, seed casings, or even roots, provided clues to their potential medicinal uses. A walnut kernel has the appearance of a human brain, and so it was concluded that walnuts should be used to treat brain complaints. Though lacking a scientific basis, the doctrine influenced medical thinking for fifteen centuries. The herbalist Nicholas Culpeper (1616–1654) was one of its last significant supporters.

Another English herbalist, John Gerard (*c.* 1545–1612), published his *Herball* in 1597. He records herbs grown in his garden in Holborn, near London, for use in remedies and cooking. Some of his herbs, such as fenugreek and nasturtium, were new arrivals from the Mediterranean and the New World.

Growing herbs for food and comfort

At first, herbs for use as food were gathered from the wild, and hunter-gatherers often developed their own ritualized ways of preparing them. In time the herbs were cultivated and farmed, and the convenience of having herbs within reach of the kitchen led in turn to the herb garden. Herbs became a staple of daily life, and in 1699 English writer and gardener John Evelyn (1620–1706) published *Acetaria: A Discourse of Sallets*, which listed seventy-three salad herbs. Herbs were used as flavorings, preservatives, preventatives of food poisoning, and even to mask the tastes of old, stale ingredients. In the West, many homes depended on their own vegetable and herb gardens until the middle of the twentieth century.

Herbs can be pungent, spicy, or sweet, and are versatile, intriguing, and surprising. No one can resist them, especially the aromatic ones, whose intense fragrances universally promote comfort, pleasure, and satisfaction. Relieving our ailments, flavoring our foods, stimulating our senses, lifting our hearts, and improving our homes, herbs enhance our lives immeasurably.

Yarrow

Although an unassuming plant, what yarrow lacks in appearance it more than makes up for with its potent healing properties. It is found growing all over the world, especially in wasteland areas, and is common in Asia, Europe, and North America. It obtained the generic name *Achillea* after the legendary Greek hero Achilles, who used the plant to cleanse and staunch the bleeding of his wounded troops in the Trojan War. Its specific name, *millefolium*, means a thousand leaves and refers to the numerous segments of its foliage. The Druids used yarrow for seasonal weather divination and as a good luck charm, which they hung over doorways to avert illness. In the ancient Chinese text of prophecy, *I Ching*, fifty straight stalks of dried yarrow of even lengths were tossed in the air and landed to foretell the future. Known as the iodine of the meadows and fields, there are references to yarrow's use in World War I as a first-aid dressing for minor wounds.

Today, yarrow is used for wildflower gardens and as an activator in compost. It is loved by butterflies, hoverflies, ladybugs, and parasitic wasps, which help keep the garden pest free. Watch your manicured lawns: if it strays it will become very invasive.

- White-pink flowers are borne in dense, flat heads and bloom from June to late fall.
- Cut back after flowering to encourage plants to bulk up.
- Cultivation is by seed or root division of the rhizomes. It requires a sunny well-drained position and is drought resistant.
- Harvest flowers and leaves as the plant comes into flower.
- The "plant doctor" of the garden, its root secretions activate the disease-resistance of nearby plants.

Yarrow has antiseptic, antimicrobial, astringent, and diuretic properties. It has been used as a medicinal herb since before medieval times and is still used today by herbalists. A tincture made from the flowers can be used to relieve rheumatic pain. Infused as a tea, it can aid digestive problems, induce perspiration, cleanse the system, and cure a cold. Use in a mouthwash for inflamed gums and oral sores; chew a fresh leaf to soothe toothache. Also infuse to make a skin wash to cleanse wounds and help prevent infection.

Yarrow has a slight aniseed flavor and was a popular vegetable in the seventeenth century. Use the young leaves only because they are less coarse. It can also be a good addition to spring soups and to white sauces.

Place a bundle in wardrobes or clothes drawers to keep away moths and other insects. In the Middle Ages, yarrow was one of the ingredients in a selection of herbs that were used to make beer, before hops.

Infuse fresh flowers for a facial steam and tonic water. Press leaves onto a shaving cut to staunch the bleeding.

Take in moderation and never for prolonged periods—it can cause skin sensitivity, headaches, and vertigo.

Yarrow flowers were used in love charms. Placed under your pillow, they would make you dream of your true love.

Thou pretty herb of Venus' tree,
Thy true name is Yarrow.
Now who my bosom friend must be,
Pray tell thou me tomorrow.

ANONYMOUS RHYME, ON SACRED EARTH WEBSITE

Sweet Flag

Native to southern Asia and central and western North America, sweet flag was brought to Europe in the thirteenth century by the Tartars who used it as a commodity and a means of bartering. It is also accredited to the Austrian botanist Clusius, who brought it back with him from Asia Minor and grew it from 1574 at the botanical gardens in Vienna. He distributed it to other botanists in Belgium, Germany, and France. The plant was introduced into Britain at the end of the sixteenth century and can be found in most parts of Britain, especially in the Fens.

Sweet flag is highly regarded in the Ayurvedic medicine of Hindu India, where it is believed to have originated and is known as "Vacha." The ancient Egyptians favored the herb in the making of perfumes. Native Americans had many uses for it; one tribe used the herb to increase strength and endurance; others used it to aid digestion and improve mental clarity.

In Europe sweet flag was often used as a popular strewing herb to ward off diseases and add a pleasing fragrance to churches and houses. At the same time, it worked as an insecticide to banish fleas.

- A vigorous, reed-like aquatic plant with sword-shaped leaves that smell of tangerine when bruised; small yellow-green flowers appear on a fleshy cane stalk. It can reach up to 5 feet (1.5 m).
- Grows in shallow water by pond edges, ditches, and marshes. Equally happy in a moisture retentive border.
- Propagate by root division in early spring or early fall.
- Harvest when the plant is at least two years old.

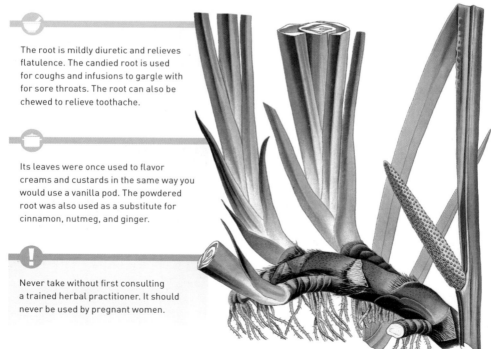

The root is mildly diuretic and relieves flatulence. The candied root is used for coughs and infusions to gargle with for sore throats. The root can also be chewed to relieve toothache.

Its leaves were once used to flavor creams and custards in the same way you would use a vanilla pod. The powdered root was also used as a substitute for cinnamon, nutmeg, and ginger.

Never take without first consulting a trained herbal practitioner. It should never be used by pregnant women.

Anise Hyssop

Anise hyssop is a striking and highly aromatic perennial herb, that attracts bees, hummingbirds, and butterflies, which seek out the tall, elegant, fluffy branching flower spikes. Native to North America, it grows wild on the prairies and plains and has traveled all the way to China.

Anise hyssop is a member of the mint family and it was widely planted by beekeepers in North America in the 1870s in order to produce a fine honey with a slight aniseed flavor.

As its name suggests it has aniseed-scented and flavored leaves and stems, which suit its folk name of "liquorice mint"; these are topped with mauve-purple flowers that bloom from July to September. There is also a pure white form of the plant, used medicinally by Native Americans.

Anise hyssop leaves can be used to flavor fish and shellfish dishes and the sauces served with them. It can be added at the end of cooking vegetables such as tomatoes, zucchini, young beets, and green beans.

The plant is antifungal, anti-inflammatory, and antiviral. Gather leaves in spring and summer and use fresh or dried to make teas. Infusions are used to relieve respiratory congestion, coughs, and colds; it increases perspiration, helping to break a fever, and improves digestion. Use as a poultice for minor burns.

The anise-scented leaves make a refreshing herbal tea. For salads, use leaves and flowers; the leaves are better chopped. It is good in egg dishes, with pork, or with strawberries or pears. Use the flowers to decorate cookies and cakes. Can be used in place of mint, especially in fruit salads and other desserts.

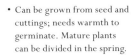

- Can be grown from seed and cuttings; needs warmth to germinate. Mature plants can be divided in the spring.
- Pinch out the flowers to promote leaf production.
- Likes well-drained rich soil in a sunny position; does not like wet conditions, which cause it to weaken and rot.
- This short-lived perennial will happily grow in a pot.

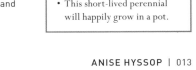

Bugle

A perennial plant native to Europe, Asia, and parts of North Africa, *Ajuga reptans* is found growing in damp woods, grassy fields, and mountain pastures. This small plant was something of a "cure-all" in the Middle Ages, but is now little used.

Commonly known as bugle, it was a favorite of the seventeenth-century English botanist and herbalist Nicholas Culpeper, who regarded it as a wound herb and a cure for hangovers. It was used to allay hallucinations resulting from excessive consumption of alcohol.

The plant may have been given the alternative name "carpenter's herb" because of its ability to staunch bleeding and heal cuts due to its tannin content. The common name is thought to derive from "bugula," a name used by apothecaries, which may in turn be a corruption of the generic name *Ajuga*. The specific name *"reptans"* means creeping and refers to the growth habit of the herb.

Bugle has a short rhizome and long leafy rooting stolons, hence its creeping habit. A short, low-growing, ground cover herb, it throws up erect four-sided stems with beautiful, dark blue-lipped flowers, which bees love as a great source of nectar.

- Bugle prefers a moist, well-drained, humus-rich soil in sun or partial shade.
- The plant blooms from May to July. Cut back after flowering.
- Can be grown from seed sown in the spring or from the runners the plant produces; these have their own root system and can reach a couple of feet or more in length.
- The scentless flowers attract bees by their coloration.

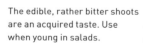

Bugle is a potent astringent used for pain relief; it was once used to lower blood pressure. In homeopathy it is used to treat sore throats and mouth ulcers. Externally, it can be applied to bruises and wounds. Cut the herb in summer to make infusions and liquid extracts.

The edible, rather bitter shoots are an acquired taste. Use when young in salads.

Only to be administered for internal use by a qualified herbal practitioner.

Alchemilla mollis

Lady's Mantle

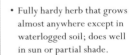

Lady's mantle is a perennial herb native to Europe, northwest Asia, Greenland, and northeastern North America; it is also found in the Himalayas. It can be found growing in damp grassland, open woodland, and on rock ledges.

The Arabic word *Alkemelych* (alchemy) was thought to be one origin of its Latin name (because of its medicinal uses). An alternative explanation is related to the leaves of the herb, which came to the attention of those seeking the mystical properties of plants; their ability to hold teardrop-shaped droplets of dew in their folds gave the plant its latin name of *Alchemilla*, meaning magical one. It's common name refers to the resemblance of the leaves to a lady's cloak (mantle)—a medieval observation. One folk tale tells that placing a leaf of the herb under your pillow will induce a "sweet slumber." Traditionally, the plant has been used for obesity, and is now thought to aid to weight loss.

The young leaves can be chopped up and added to salads or vegetables.

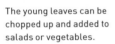

Infused dried leaves are used as an astringent and facial steam for acne. Make a cold infusion to use for a compress on puffy eyes.

Medicinally the plant is held in high esteem. It is known for its use in treating menstrual problems and for strengthening and healing after childbirth. Its pain relief qualities come from the action of the salicylic acid contained within the plant. Use dried leaves to prepare a mouthwash, or a poultice for healing wounds.

Due to its tannic properties, it will produce a bright green dye for wool.

To be avoided during pregnancy or when breast-feeding.

- Fully hardy herb that grows almost anywhere except in waterlogged soil; does well in sun or partial shade.
- Can be grown from seed or division. Cut back hard after flowering to encourage new growth.
- Excellent edging and ground cover plant.
- Self-seeds freely and can become invasive.

Garlic

Allium sativum, commonly known as garlic, has been cultivated for around 6,000 years and came originally from central Asia. A species in the onion genus, it has been associated throughout history with promoting strength and health and has been held in high esteem in the realms of religion and magic, too. This pungent, hardy perennial has also been widely cultivated across the centuries as one of the most useful culinary herbs. In ancient Egypt, it was fed to the workers building the pyramids to give them strength and protect them from illness and epidemics. Greek athletes and wrestlers would chew a few cloves of garlic to give themselves strength and courage. They also used it as currency, such was its value and regard. The Roman physicians Galen and Dioscorides thought it the great panacea for the common man and recommended it for many ailments. The ancient Greeks left cloves of garlic on crossroads as a food for Hecate, their goddess of magic, witchcraft, the night, and the moon. Garlic was also used as a remedy against the plague during the Middle Ages. Garlic has always carried the mantle of a strong aphrodisiac; men of the cloth were therefore forbidden from its consumption.

- Easily grown, it prefers a sunny position in well-drained soil. Add compost and sand to help bulbs establish comfortably.
- Not happy in acid soils; try growing in deep containers.
- Plant in fall/early winter for best results. Traditionally planted on the shortest day and harvested on the longest. It is usually ready when the top growth changes color and keels over.
- Hang up to store over winter in a cool, dark place.

Garlic has a strong penetrating flavor; use sparingly so as not to overpower the food you are serving. For a hint in salad or a fondue, rub the bowl with a bruised clove. Infuse whole cloves of garlic in bottles of olive oil and wine vinegar. Garlic can enhance chicken, lamb, bean soups, and casseroles; it is also good in pasta, pizza, passata, pesto, hummus, and aioli. Use garlic scapes (flower stems) in salad or gently wilted as a vegetable. To counteract garlic breath, chew fresh parsley, celery leaves, or cardamom seeds.

Although garlic is generally commended for its beneficial qualities, care should also be taken when used medicinally. Not to be taken in large quantities by nursing mothers or those suffering from eczema.

Hanging bunches of garlic around the home was thought to keep evil spirits, vampires, and trolls at bay. Brides-to-be were "beaten" with garlic stalks to protect them from illness and ensure that they bore healthy children.

Garlic has powerful medicinal properties: antibacterial, antibiotic, antifungal, antimicrobial, antioxidant, antiparasitic, antiseptic, and antiviral. For garlic to be really effective it needs to be eaten raw; start slowly with one clove a day, building up to three; add it to raw foods. To treat the fungal condition *Candida albicans*, it kills off excess bacteria while establishing healthy gut flora. Eating raw garlic will keep mosquitoes and other biting insects away. Macerate the cloves in honey to treat sore throats, coughs, and colds.

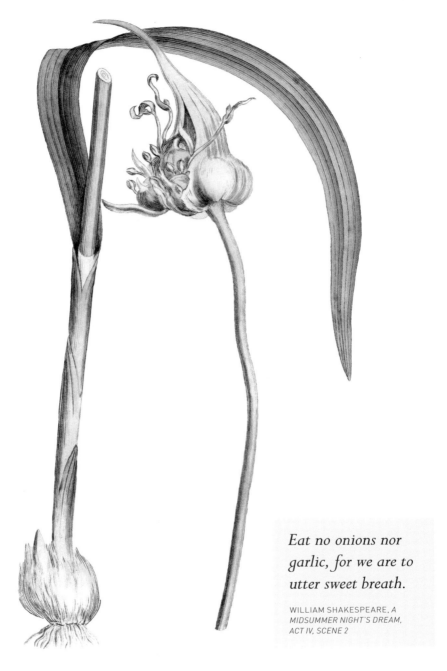

*Eat no onions nor
garlic, for we are to
utter sweet breath.*

Allium schoenoprasum

Chives

Records show that chives were in use 4,000 years ago in China. The explorer and traveler Marco Polo was the first to extol their culinary worth to the West, where they soon became indispensable. In the Middle Ages, chives were thought to cure melancholy and were hung in dried bunches around a home in order to drive away evil spirits. They were not cultivated in European gardens until the sixteenth century. Today they can be found growing wild all over Europe, Asia, and North America. A bulbous perennial plant with a distinctive onion or garlic scent, it is the smallest species of the edible *Allium* genus. Gardeners often favor it for the pale purple to pink and pure white bell-shaped flowers borne in umbels that top the aromatic stems in summer. It is a very useful edging and hedging herb around vegetable and decorative gardens due to its ability to repel certain pests.

The leaves, bulbs, and flowers of chives can be eaten and are especially good with potatoes and eggs. Use the leaves and bulbs to garnish and flavor soups, salads, soft cheeses, omelettes, and sauces, such as rémoulade and ravigote. Use the flowers in salads and herb butters. Brings flavor and color to crème fraiche and sour cream. Add to cooked dishes when serving because the delicate flavor diminishes with heat.

- Clump-forming perennial with slender, cylindrical, hollow leaves.
- Cut back after flowering and feed to encourage new leaf growth.
- Split every third year in the fall to prevent compaction and to encourage healthy new growth. Can be grown from seed and division of root stock.
- Likes rich, moist soil in a warm, sunny position but will tolerate shade. Do not allow to dry out in summer.

Chives contain iron, calcium, pungent, volatile oils, pectin, and small quantities of sulfur. Mildly antibiotic, they have a reputation for strengthening the stomach and reducing high blood pressure.

Planting chives by carrots and parsley seems to improve their health and flavor. A strong infusion of chives is used as an antifungal spray on vegetables for both downy and powdery mildew.

Aloe Vera

Aloe vera is a tender, evergreen, perennial plant, which was first described by botanist Carl Linnaeus in 1753. It is native to Africa, especially the eastern and southern regions, where it can be found growing wild. It is also found in the West Indies and other tropical areas and has been reported growing in the Zapata area of Texas. It is also a popular household plant. Common names include Chinese aloe, Indian aloe, and true aloe. The word "vera" means true or genuine.

The plant has been used in herbal medicine since ancient times for its healing and soothing qualities. It is an embalming agent—Jesus's body was wrapped in linen impregnated with myrrh and aloes. Ancient Egyptian wall paintings depict aloe vera as a revered healing and beauty enhancing plant; Cleopatra's beauty was partly attributed to the use of aloe vera. Native Americans called it "the wand of heaven" and used it to treat desert sunburn and scorpion bites. To this day, aloe vera is used in cosmetics, hand creams, and suntan lotions, and more recently it has been used to treat radiation burns.

- A succulent with strong, brown, fibrous roots; its stemless plant base produces tapering pointed blades, often with spiny teeth along the margin. If and when it flowers, they are narrow, trumpet-shaped, and yellow or orange in color. Plants need constant warmth to flower.
- Propagation is by offsets that form at the base of the plant. Aloes are sensitive to temperature fluctuation and need to be kept above 50°F (10°C).
- Do not over water or they will rot.
- Plants of older than two years have stronger potency.

Aloe gel soothes dehydrated skin and helps to diminish scars. Treat sunburn by splitting a leaf and wrapping around the affected area.

Always seek the advice of a medical herbalist before use, especially aloe latex (outer lining of leaves).

The centers of aloe leaves contain a clear mucilaginous gel that has anti-inflammatory and healing properties. Best known for its treatment of burns, it can be applied directly; when applied to wounds, it forms a clear protective seal, which encourages skin regeneration. In recent years, commercial preparations have been used to treat digestive problems, including constipation.

Aloysia citrodora

Lemon Verbena

Lemon verbena is a highly aromatic deciduous shrub that is native to Chile and Peru. It was brought to Europe in the seventeenth century by the Spanish, who cultivated it for its perfumed oil. The genus name *"Aloysia"* refers to Maria Luisa, Princess of Parma, who was the wife of King Charles IV of Spain. This connection gives the herb its Spanish name "Yerba Luisa." The herb stands out from the crowd by virtue of its sharp, lemony scent and flavor, which is floral and zesty.

The elegant, elongated leaves are bright green and pliable when they first unfurl, but as the plant matures they become rough in texture and slightly tough. The flowers are small, delicate, and typical of the Verbenaceae family, but with an intense perfume of their own. Lemon verbena is a very tactile plant: it is hard to resist picking a leaf, crushing it between one's fingers, and then inhaling its delightful scent. Ideal placement for the plant is by the back door, where it can be brushed against as you walk past. Lemon verbena has culinary uses like flavoring tea and cakes. It is also especially good for repelling gnats and other insects.

The essential oil is used in cologne, perfume, and soap. Add a handful of leaves to infuse in bathwater to sooth and relax. Macerate the leaves in almond oil for massaging; add this oil to homemade lotions.

Lemon verbena is antispasmodic and is used to treat digestive disorders. It also aids digestion and relieves indigestion and flatulence. A tisane before bed is soothing and relaxing; use for anxiety, nervous tension, and stress.

Leaves can be used fresh or dried in cakes, buns, tea breads, and scones. Lemon verbena is particularly good in desserts, sorbets, ice cream, custards, curds, compotes, and tarts. Do not overheat—this diminishes the intensity of the herb. It can also be used in savory dishes as a substitute for lemongrass, and is good in liqueurs and syrups.

Lemon verbena should not be taken in the long term because it can become a stomach irritant.

- It can be grown from seed, tips, and cuttings.
- The fresh cuttings have a tendency to wilt quickly, so be prepared for this and move them from plant to cutting pot or bed immediately.
- Keep young plants in pots for the first two years, then plant outside when warmer weather appears in well-drained, sunny position.
- Always cut back after flowering and give a feed.

Marsh Mallow

Marsh mallow is a perennial herb, native to Britain, Europe, North Africa, and central Asia and it has also been introduced to North America. It was once found growing wild in ditches, near the sea, and on salt marshes. It was first recorded for its healing properties in the ninth century BCE by the Greeks; the generic name *"Althaea"* originates from the Greek word "altho," meaning "to cure." The specific name *"officinalis"* refers to its long history of use as a medicinal plant.

Romans, Syrians, Greeks, and Armenians considered it a delicacy and often subsisted for weeks on marsh mallow, which at that time was in abundance when other crops failed. In France, the tender young growth of the plant was eaten raw in salads as a spring tonic. A confection made from the root since ancient times has evolved into today's sweet treat.

The fresh seeds, known as "cheeses," can be sprinkled on salads. Use whole flowers to decorate a salad. Eat young leaves in salads or add them to oils and vinegars. Steam leaves as a vegetable. Roots need to be boiled first to soften them, then sautéed in butter.

Marsh mallow has anti-inflammatory and antibacterial properties. Due to its mucilaginous nature, it has been used to sooth bronchitis, whooping cough, and digestive problems. Externally it is used for skin inflammation, insect bites, and splinters. The pulverized root is made into a soft lozenge for sore throats, from which today's confection originates.

- Marsh mallow is a hardy herbaceous perennial with gray-green leaves that are velvety on both sides.
- Prefers moist moderately fertile soil in full sun.
- Sow seed in spring or divide cultivated plants or stem cuttings from non-flowering stems.
- Pale pink flowers bloom in August and September.

Leaves and roots are used to make soothing lotions and potions for dry hands, sunburn, dry brittle hair, and face masks.

The emulsifying properties of the roots are used to clean Persian rugs in the Middle East, as they preserve the color of the vegetable dyes.

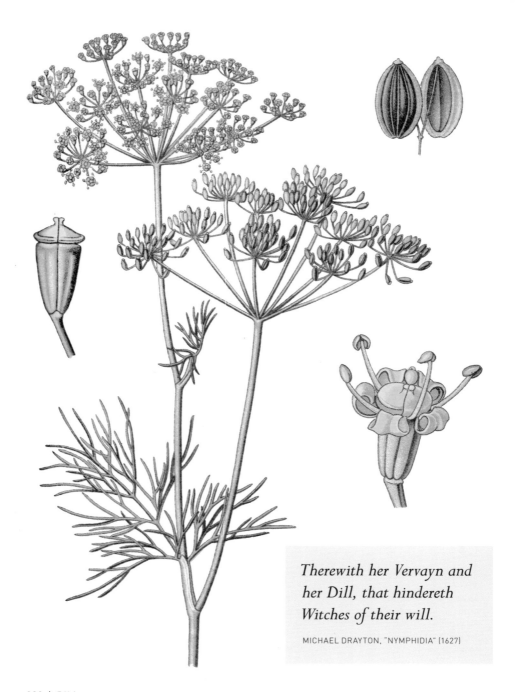

Therewith her Vervayn and
her Dill, that hindereth
Witches of their will.

MICHAEL DRAYTON, "NYMPHIDIA" (1627)

Dill

Dill is a hardy annual that is native to eastern Europe, western Africa, and the Mediterranean region. It has also long been cultivated in India for both medicinal and culinary purposes. The word "dill" is said to come from the Anglo-Saxon "dylle" or the Norse "dilla," which means to soothe or lull and refers to its traditional use as a remedy for indigestion.

The herb was much prized by the ancient Greeks who regarded it as a sign of wealth and used it to treat hiccups. The Scythians used it to embalm their dead; twigs of dill were also found in the tombs of the pharaohs in ancient Egypt. In Rome, gladiators would be fed meals covered with dill, as it was thought to give them courage. Dill was one of the nine sacred herbs used in pagan festivals. It was later a magicians herb and protector against witchcraft during the Middle Ages, while mere mortals infused it in wine to arouse passion. Early settlers took dill to North America, where it became known as "meeting house seeds" because it was nibbled during long church sermons to prevent the pangs of hunger.

Interestingly, like many other plants in the Apiaceae (Umbelliferae) family, the essential oils found in dill leaves differ from those in the seed, with the former being used in food and the latter in cosmetics.

- An elegant umbellifer with delicate highly pungent fern-like leaves.
- Sow seeds successively from early spring to late summer to ensure a constant supply of leaves.
- Sow into module/plug trays so as not to disturb the new roots, or when there is no threat of frost in open ground.
- Prefers well-drained good rich soil in full sun; if poor soil, protect from the wind.
- Keep it well away from fennel, or they cross-pollinate and mix up individual flavors.

Seeds are used in traditional pickling spice, such as that for pickled cucumber, and certain curries call for dill seeds. Add the leaves to freshly cooked broad beans, mayonnaise, sour cream, crème fraîche—all of which benefit from a few snipped leaves of dill. Always add fresh leaves just before a dish is served for optimum flavor. Gravlax—a Nordic dish of raw salmon—is cured in salt, sugar, and dill. It is also good with sole, halibut, and trout. Try it with avocado and sour cream. Snip with scissors, don't chop!

Dill is a very good source of calcium and can be taken as a nail strengthener. Chewing the aromatic seeds freshens one's breath. Hippocrates, the father of medicine, used it in a recipe for cleaning the mouth.

Dill is excellent when used as companion planting. As its umbrella flower heads go to seed, they attract many beneficial insects, some of which like to feast on aphids.

Dill has been revered since ancient times for its numerous healing properties: it is antispasmodic, digestive, calmative, and a relaxant. The seeds are good for digestive disorders, bloating, colic, and gas. It is especially good for use with babies; dill is an ingredient of gripe water. An infusion of dill seeds can be made into a tea to treat hiccups. It can also be taken in the same way to relieve insomnia, because it aids relaxation and induces sleep.

Angelica

Angelica is an attractive umbelliferous plant. Majestic, statuesque, and sweet-scented, the herb's virtues have been extolled for centuries for its curative properties and unique flavoring. The seventeenth-century botanist Nicholas Culpeper recommended it for treating everything under the sun. Angelica grows wild in the northern parts of Scandinavia, Iceland, Greenland, and the Faroe Islands. It is cultivated in France, Germany, Romania, and some Southeast Asian countries, such as Thailand. It is also commonly known as holy ghost, wild celery, and Norwegian angelica.

The herb has long been associated with protection and healing. One legend refers to its name deriving from an angel who revealed in a dream to a monk that the herb could cure plague. The Swiss German Renaissance physician Paracelsus described it as a "marvelous medicine" during the plague epidemic of 1510 in Milan. Another legend tells of angelica blooming on the feast day (May 8 in the old Julian calendar) of Michael the Archangel, the great defender who protects against evil. Laplanders crowned their poets with its leaves and flowers, believing that the scent of the plant inspired them. In the language of flowers, it is said to be a symbol of ecstasy, inspiration, and magic.

- Angelica can be biennial or a short-lived perennial, living up to four years.
- Remove the emerging flowers to prolong the life of the plant. Once it sets seed, it dies.
- Seeds need to be sown fresh as the viability of the seed declines rapidly.
- Often any seeds that fall to its feet will germinate in the hundreds.
- The plant likes humus-rich, moisture-retentive soil with a dappled shade.
- Bees favor its flowers.

Angelica has been cultivated as a vegetable for centuries. Use its young tender leaves in salads with other herbs and leaves. Cook the stems with gooseberries, rhubarb, and other tart fruits to reduce the acidity. The stems are also candied for food decoration. The seeds are used to flavor liqueurs, such as Chartreuse and Bénédictine. Angelica cordial was a seventeenth- and eighteenth-century tipple made from the roots. Make a refreshing tea using the young leaves, which taste like China tea.

Angelica is antifungal, anti-inflammatory, antibacterial, digestive, diuretic, and an expectorant. It stimulates circulation and is good for chronic fatigue. It is helpful in treating cystitis and also nervous headaches and anemia. For sore throats, gargle with an infusion made from the leaves. Harvest stems in April or May and the leaves in May and June before flowering. Gather roots at the end of the fall, for drying and storing; they will keep their medicinal properties for years. Use powdered root to treat athlete's foot.

Angelica has a pervading aroma, quite different from other members of its family, such as fennel, anise, and caraway. Add leaves to warm bathwater to soothe and relax.

Angelica archangelica is not to be given to diabetics. It must not be taken in large doses. Take care if applying directly to the skin in sunlight. Avoid during pregnancy and breast-feeding.

In England, where it was also known as bellyache root, dried angelica roots were made into powder and mixed into wine to "abate the rage of lust in young persons."

PINKIE D'CRUZ, ON ANGELFIRE WEBSITE

Anthriscus cerefolium

Chervil

Almost everything about chervil is delicate—its appearance, growth, and flavor, which is sweetly aromatic with an aniseed/parsley taste. The plant is known as one of the aristocrats of the herb world. It is native to the Caucasus and was distributed by the Romans throughout most of Europe, where it became naturalized. Chervil was cultivated in Brazil in 1647. However, there is no mention of its use in America until a century later.

Chervil has had many names throughout history, including garden chervil and French parsley. Linnaeus gave it its present name, *Anthriscus cerefolium*. "Ant" comes from the Greek word for flower and "cerefolium" means "leaf of ceres," referring to Ceres, the Roman goddess of agriculture, who was thought to be responsible for the fertility of the land. Chervil was a popular herb at Lent and was traditionally eaten on Maundy Thursday for its blood cleansing properties and also for being uplifting and restorative. Chervil seeds were one of the ancient Egyptian funerary herbs and were found in the tomb of Tutankhamen. The leaves of the dried herb were once burned to comfort the bereaved and to help them contact the deceased. Chervil was also one of the nine sacred herbs of the Anglo Saxons.

- Hardy annual
- The distinctive seeds are sharp, slender, and elongated. Seed is sown in early spring, straight into open ground or plugs. Hates root disturbance.
- If the seed is fresh, it will be quick to germinate with the right weather conditions and light levels.
- Likes moisture-retentive soil. Prefers semi shade and dislikes full sun. It will bolt; if this happens it will alter the flavor.
- Often self seeds.

Chervil, like most aromatic herbs, is used as a companion plant to enhance not only the flavor and fragrance of its neighbors but also their growing vigor. In particular, with chervil growing alongside, radishes will be crisper and hotter, and lettuce and broccoli improved in taste and more substantial.

Chervil is one of the traditional *fines herbes* of French cuisine. It is used extensively in France in omelettes, salads, and soups. The edible roots are eaten as a vegetable or added to soups and stews. It is good with chicken, white fish, fresh asparagus, and chopped and sprinkled on freshly cooked broad beans, seafood, peas, and young carrots. Never overheat chervil because its color and flavor will be destroyed; always add at the end before serving a dish. Fresh chervil leaves are available throughout the winter months.

Chervil has had various uses in herbal medicine: it has been taken as an antidepressant, anti-inflammatory, and a digestive. Taken as an infusion, it makes a good spring tonic for detoxing—it cleanses, stimulates digestion, and lifts the mood. It has been used to treat chronic catarrh and sinusitis. Traditionally, it was used for bad dreams, burns, and stomach upsets. It has a high calcium content.

Use the leaves infused or freshly juiced to make a natural skin cleanser and tonic. The plant also has healing properties and is used to treat eczema.

<blockquote>
Sweet chervil is so like in taste unto Anis seede that it much delighteth the taste among other herbs in a sallet [salad].

JOHN PARKINSON, *PARADISI IN SOLE PARADISUS TERRESTRIS* (1629)
</blockquote>

*The Horse Radish . . . is commonly used
among the Germans for sauce to eate fish
with and such like meates as we do mustarde.*

JOHN GERARD, *THE HERBALL, OR GENERALL HISTORIE OF PLANTES* (1597)

Armoracia rusticana

Horseradish

A perennial plant of the *Brassicaceae* family, which includes mustard and wasabi, horseradish is grown mainly for its large, white, tapered root. It has been cultivated from the earliest times and is a popular culinary herb and condiment all around the world. The root has little aroma, until cut or grated, when it has a pungent odor that can irritate the eyes and sinuses. Both the roots and leaves were used medicinally by the ancient Greeks and Romans, and throughout the Middle Ages.

The plant is thought to have originated in eastern Europe from the Caspian, through Russia and Poland to Finland. It is said to be one of the five bitter herbs that the Jewish people were required to eat during the feast of the Passover. Immigrants living in Illinois in the mid-1800s, planted horseradish with the sole intention of selling the roots on a commercial basis. To this day Collinsville, Illinois, is referred to as the "horseradish capital of the world."

Escaping from cultivation, horseradish can be found growing wild all over Europe, the British Isles, North America, and New Zealand. The "horse" prefix refers to the strength, size, or coarseness of the plant.

- A fully hardy perennial, horseradish prefers well dug, rich, moisture-retentive soil if you wish to produce good large root stocks.
- Propagation is by root division, root cuttings, or seed. Divide mature root crowns in spring.
- Dig roots in the fall when plumped up for winter. Store roots in sand.
- Left undisturbed, it can spread via underground shoots and become invasive.

Grow near a potato patch to keep away certain diseases and produce a healthier crop. A horseradish tea used for treating mildew on plants can be made from the leaves. Steep the leaves for forty-eight hours in 1 ¾ pints (1 litre) of water for maximum strength, then dilute four parts water to one part horseradish solution. Boiling the leaves will give a deep yellow dye.

Horseradish is antioxidant and antibacterial. It has been found to have natural antibiotic properties. In herbal medicine, it has been used for gout, rheumatic and lung conditions, digestive problems, and urinary infections. Externally it can be used as a poultice for arthritis, rheumatism, and sciatica. It must not be taken if you have sensitive skin.

Young leaves can be used in salads and to wrap whole fish, such as trout, before baking or steaming it. To make horseradish sauce, the root is first scrubbed then grated; it is excellent mixed with beets and sour cream. It can also be pickled and used to make a vinegar. Make preserved horseradish by thinly slicing the root, drying it on a tray very slowly in a warm oven, then pound in a pestle and mortar and store in an airtight jar. It will keep its strength for many months. Always add a little lemon juice when grating; it fixes the taste, which can deteriorate quickly.

Not to be used by those suffering from thyroid problems, stomach ulcers, and kidney problems. Once the plant is established, use it, otherwise it will become invasive in the garden. Can be grown in a container if it is deep enough.

Arnica

The name "arnica" is a derivative of the Greek word "arnakis," meaning lamb's skin, and is a reference to the leaves, which have a soft woolly texture. It is also known as sneezewort because the freshly crushed flowers cause sneezing. Another common name, "tumblers cure," comes from its well-documented external use as a soothing remedy for bruises and sprains. The leaves and roots were once smoked as herbal tobacco, hence its folk name "mountain tobacco." It is also sometimes known as wolf's bane and mountain arnica.

Native to central Europe, arnica is endemic to Europe with the exception of the British Isles, and the Italian and Balkan Peninsulas. It was once found in Scandinavia and parts of North America, where other species with similar properties may be found. Historically arnica is mentioned in numerous world pharmacopoeias and in U.S. editions up to the early twentieth century for treating numerous ailments.

Once found in abundance in Alpine meadows, it is now scarce in the wild, mainly because of increasingly intensive agriculture, and is protected.

- Hardy perennial plant that likes sandy acid soil in a sunny position.
- Can be grown from fresh seed. Sow in late summer/early fall. Place in a cold frame; needs to be kept cool because heat stops germination; slow to germinate taking up to two years.
- Divide plants in spring once they appear.
- Large scented yellow flowers bloom throughout summer, from June to August.

Not to be taken internally and never apply to broken skin. Arnica is subject to legal restrictions in some countries.

Arnica is anti-inflammatory. Its pain-relieving properties have been used to treat sore and aching muscles and rheumatic joint problems. It has long been taken as a homeopathic remedy for shock and trauma and has been known as the "accident herb." Flowers can be made into a tincture or applied in the form of a poultice, compress, oil, or salve for muscular pains, sprains, chilblains, and joint pain.

Southernwood

Southernwood is a strongly aromatic shrubby member of the *Artemisia* genus. It is found growing in southern, eastern, and central Europe, as well as parts of North America, but it rarely grows in the wild.

The generic name is said to come from either the ancient Greek goddess Artemis or the sister of the Persian king Mausolus (377–353 BCE), Artemisia, who was renowned for her botanical and medical knowledge. The old English name of "southernwood" refers to its southern European origins as a "wood from the south." The English herbalist Nicholas Culpeper is known to have used the plant for all sorts of ailments, from sciatica to baldness. It was also once used to ward off infection and was one of the aromatic herbs used in nosegays up until the nineteenth century.

The sweetest of all the perennial *Artemisia* species, southernwood makes an unusual aromatic hedge because its growth pattern makes it easy to manage.

The lemony leaves can be used sparingly to flavor some cakes. In Italian cuisine, it is used together with other herbs to roast lamb. It can be added in small quantities to salads, aromatic vinegars, and certain stuffings for game.

Dried leaves are used in potpourri. Fresh and dried bunches are hung in wardrobes to prevent moths. It is used in sachets to repel fleas; fresh leaves combat flies and mosquitoes.

A sign of affection and fidelity, sweetly scented, lemony southernwood was used in country bouquets exchanged by lovers.

Used as a tonic tea and as a vermifuge for threadworms. Externally, use for swellings, frostbite, and hair loss.

Not to be used by pregnant women due to its ability to restore menstrual flow.

- Deciduous or semi-evergreen shrub; a hardy perennial grown for its aromatic foliage. Needs a hot climate for it to flower.
- Woody with gray-green finely divided leaves.
- Prefers a fully sunny position with light, well-drained soil.
- Propagation is best by softwood cuttings in spring. Cut back mature plants in spring after the frost, never in winter.
- Harvest leaves for drying in midsummer.

You never know what that little green fairy will do.

Wormwood

With the exception of rue, wormwood is the bitterest of all herbs. It stands above all others in the *Artemisia* genus as the most bitter, hence its specific name, which comes from the Latin *absinthium*, meaning "without sweetness." Its common name refers to its use as a vermifuge or expeller of worms. Wormwood can be found growing wild in Britain and Europe; it is also found in Siberia, Eurasia, North and South America, and New Zealand. It is also known as absinthe sagewort.

Ancient Greeks and Egyptians valued wormwood for its stimulating medicinal properties. For many centuries, Native Americans have used wormwood for a wide range of ailments and as an insecticide, especially for bed bugs. The Chippewa tribe boiled the plant tops, which they used in a warm compress for sprains and strained muscles.

Despite being known for its beneficial uses, excessive use of wormwood can be harmful, addictive, and poisonous. It is a hallucinogen and can damage the central nervous system. Absinthe, the French alcoholic drink extracted from the root, was known to have killed the French poet Paul Verlaine and the artist Toulouse Lautrec. It was made illegal in 1915 in the United States and most of Europe.

After World War II, it was grown commercially in the state of Michigan for use in various liniments. Although it is no longer grown there, it is still used for treating sprains, especially in horses.

- Hardy perennial, partial evergreen.
- Likes full sun and prefers well-drained light soil. Can be grown successfully from seed or softwood cuttings from new growth in early summer.
- Divide every three years to keep the plant healthy and in check. Spring division is the most successful.
- Prune back in fall.
- Protect when temperatures may fall below 23°F (-5°C)
- Artemisia, especially the silver forms, make good garden foliage plants in the herbaceous border.

Due to its strength of aroma and toxic root excretions, wormwood is not compatible growing with a number of plants, including dill, cilantro, fennel, and sage. However, its toxic emissions have been known to help guard fruit trees against caterpillars, moths, and aphids. Supposedly a tea of wormwood sprinkled on the ground will deter slugs and mice. It has also been used to make a spray against whitefly, thrips, and red spider mite.

Wormwood is antiseptic, a cardiac stimulant, and a relaxant. It has been used to treat digestive problems; it can stimulate the appetite and relieve indigestion. It also reduces flatulence and expels worms, especially roundworm and threadworms. Its stimulant properties tone the uterus, liver, and gall bladder and increase bile. Externally, it can be made into a compress to treat bruises and sprains.

Wormwood contains a psychoactive chemical that can cause kidney failure. Only take internally under the supervision of a qualified medicinal herbalist.

Inhaling wormwood was supposed to increase psychic powers. It was used for astral projection, inducing visions, and divination, and included in love potions.

Tarragon

Tarragon has been in use since 500 BCE, when its use was recorded by the ancient Greeks, who considered it one of the "simples"—remedies that use only one herb. It was also recommended by Hippocrates, the father of medicine.

Tarragon is a prized culinary herb, particularly in French cuisine. It also features in the cuisines of southern European countries, southern Asia, and Siberia. A native of southeastern Russia, it appeared in North America with the early settlers. It was only grown in the royal gardens of England from the fifteenth century onward; by the sixteenth century it was in more common use.

The Russian form *Artemisia dracunculus dracunculoides* has a more strident habit, but the leaves have much less flavor. Its specific name means "little dragon," which may have derived from tarragon's fiery tang or from its serpentlike roots. Being known as "the dragon herb" probably led to the belief that it cured the bites of venomous creatures. The herbal doctrine of signatures refers to the "roots serpent like cure bites of venomous snakes, insects, and mad dogs"; the taste is recorded as "hot and biting."

Many Native American tribes used it for a host of ailments, such as the Hopi, who boiled or roasted it between stones. It was also used by Pawnees to make of mats, rugs, and bedding.

- A perennial that rarely produces viable seed and only in warm climates.
- Propagate by cuttings or division of the roots after the frosts. Pull the roots apart, do not cut.
- French tarragon has a finer growth and needs a warm well-drained position; dislikes wet cold soil. Replace every three years as the flavor deteriorates as the plant matures.
- Keep dry throughout winter and frost-free.
- Russian tarragon, known as "false tarragon," can be grown from seed.

French tarragon is used in many dishes for its intense aniseed flavor. It is especially good with chicken, pork, rabbit, egg dishes, and tomatoes. It is one of the *fines herbes* of French cuisine, and one of the components of *herbes de Provence* and Dijon mustard. Classic French sauces, such as Béarnaise and rémoulade, rely on tarragon for their distinctive flavors. It is used in vinegars, herb butters, and soups, and with fish such as halibut, sole, and scallops.

Tarragon is a digestive, sedative, and carminative herb. Infusions can be taken to stimulate poor digestion and treat bloating, flatulence, nausea, and constipation; it is also effective for worms in children. The mild sedative effect is used to relieve insomnia. Historically, it was used for toothache and as a breath freshener. Tarragon has mild menstruation-inducing properties and is used if periods are delayed and for this reason it is not recommended in pregnancy.

Tarragon seems to invigorate those plants growing nearby, especially sweet peppers and eggplant. Secretions from the roots repel damaging nematodes.

Add tarragon to a complete fresh fruit salad, or sprinkle on ripe melon, peaches, nectarines, or apricots. It adds piquancy to lime- or lemon-based desserts.

'Tis highly cordial and friendly to the head, heart, and liver.

JOHN EVELYN (1666)

Atriplex hortensis var. *rubra*

Red Orach

Orach is a tall, erect decorative hardy annual, also known as mountain spinach. It was once a common sight in vegetable gardens, where it grew as a substitute for spinach.

Thought to have originated in eastern Europe, it can still be found growing throughout Europe and the British Isles. The red form, known as red garden orach, has gained in popularity and is much loved throughout Continental Europe, where it is eaten as a vegetable and salad herb. In Romania it features in a spring herb soup that is served cold. In antiquity in Mediterranean regions people ate orach as a regular herbal vegetable. It was widely grown from the Middle Ages up to the nineteenth century, then fell out of favor, although it is regaining popularity.

Red orach is also known as goose foot, which comes from the appearance of the leaves, and saltbush because of its ability to tolerate salty and alkaline soils. It makes a great addition to the herb garden and, if allowed to reach its full potential, is a dramatic plant for the flower border with its plumes of flat fruits, the rubra being the most attractive of the genus.

- Hardy annual with variously shaped leaves that reaches a height of 4 feet (1.2 m).
- Sow the disklike seed for early production in plug trays under protection or sow into open ground after the frosts for quick germination.
- Keep pinching out any flowering shoots to allow the leaves to proliferate.
- Allow one or two plants to run to seed and they will invariably self-sow; or collect seed, which remains viable for three years.

Red orach has a spinachlike flavor. Use the young smaller leaves in salads and the larger leaves like spinach. It can be cooked in soups, purees, and vegetarian casseroles. Add right at the end to stir fries. It can be used to balance the acidic flavor of sorrel. Rich in calcium, phosphorus, magnesium, iron, vitamin C, and carotene, it is a healthy alternative to other salad greens and it also makes healthy rich smoothies. Red orach retains salt.

Red orach has laxative, diuretic, and purifying properties. It aids digestion, cleanses the kidneys, and supports the gall bladder. The seeds were once mixed with wine to induce vomiting and cured the ailment known as "yellow jaundice." It was also applied as a poultice for boils and swellings. It is no longer favored for medicinal use because other herbs are deemed more beneficial.

Deadly Nightshade

Native to the British Isles, deadly nightshade is found growing throughout Europe and has become naturalized in the United States. A perennial of the potato family (Solanaceae), it was once an important crop in Croatia and Hungary before World War I. It is usually found growing in old ruins and waste ground, creeping and twining through the undergrowth. The foliage and berries of the plant are highly poisonous and can cause delirium and hallucinations. The drug atropine is derived from it.

For this reason, deadly nightshade is a plant with a sinister past. It has a long history as a medicine and poison and has been mentioned historically in a number of unpleasant circumstances. The common name comes from sixteenth- and seventeenth-century apothecaries, who named it "solatrum mortale." Botanist Pietro Andrea Mattioli noted that it was the Venetians who first called the plant "herba bella donna" ("beautiful woman herb"). Venetian ladies used a diluted water of the herb as an eye-drop to dilate and enlarge their pupils to appear more attractive. It is still used today for this purpose by ophthalmic surgeons to help them examine the eye.

- Self-seeds freely.
- The plant is wild but it has been grown commercially as a crop, for example in the United Kingdom in Hertfordshire and Suffolk. The leaves are cropped for the two component alkaloids, atropine and hyoscyamine, both of which are used in the medical profession.
- Strongly recommended that this plant is not cultivated in any garden because of its highly poisonous nature.

Maud Grieve wrote in her book *A Modern Herbal* (1931) that the generic name, atropa, comes from Greek mythology. Atropos was one of the three goddesses of fate and destiny who held the shears to cut the thread of human life, referring to its deadly poisonous nature.

This is one of the herbs of the "flying ointment" of German witches, along with soot and aconite, henbane, and hemlock. Rubbed onto the skin, this was said to enable them to fly astrally.

All parts of the plant are extremely poisonous and only to be used under strict medical supervision. *A. belladonna* adversely affects the brain, heart, and bladder and causes delirium and hallucinations before leading to unconsciousness and sometimes death.

I noticed on the margin of a pool
Blue-flowering borage, the Aleppo sort,
Aboundeth, very nitrous. It is strange!

ROBERT BROWNING, "AN EPISTLE CONTAINING THE
STRANGE MEDICAL EXPERIENCE OF KARSHISH,
THE ARAB PHYSICIAN" (1855)

Borago officinalis

Borage

Indigenous to countries around the Mediterranean, and now found growing wild in northern Europe also, borage reached North America by means of early settlers who valued the plant and its beautiful five-petaled flowers enough to carry the seeds with them.

Since ancient times, borage has been used for uplifting the spirits. The herbalist John Gerard noted in *The Herball, or Generall Historie of Plants* (1597) that: "Those of our time do use the flowers in salads to exhilarate and make the mind glad. . . . The leaves and flowers of Borage put into wine make men and women glad and merry, driving away all sadness, dullness, and melancholy."

The genus name *Borago* may derive from the Latin *borra*, meaning "rough hair," a reference to the hairiness of the leaves and stalks, or *corago*, which translates as "I bring courage to the heart." It may also be associated with the Celtic word *barrach*, meaning "man of courage." The flowers of borage were among several floral motifs that the ladies of knights fighting in the Crusades would sew onto their menfolk's clothing to fortify them in battle. Before battle, the knights themselves would consume drinks with borage flowers floating in them for the same purpose—as, possibly, did the pioneering American settlers.

- Borage grows well from an early spring sowing of seed.
- A hairy annual with upright, hollow stems, borage prefers a free-draining soil and a sunny position.
- Borage will tolerate poor soil but the plant will attain better proportions if conditions are favorable.
- The plant produces many star-shaped, azure-blue flowers. Deadhead plants to prolong flowering.
- It will naturally self-seed.
- Due to its large taproot, it will not transplant easily.

Borage leaves and flowers were once used in the manufacture of the beverage Pimm's, before being replaced by mint. They are still used to garnish the drink.

Borage seeds are used to make starflower oil, a product used by herbalists as an alternative to evening primrose oil. The oil is also used in aromatherapy as a base oil, diluting powerful essential oils, because it is safe and healing on the skin.

Bunches of the fresh herb were once hung in bedrooms and wardrobes for their sweet scent. Fresh leaves and flowers may be added to bathwater for their soothing and relaxing effect; they will also help to moisturize dry skin, and soothe and heal any skin inflammation.

Borage tea is used to treat such digestive disturbances as gastritis and irritable bowel syndrome. It may also help to cure a hangover.

Borage contains twice as much gamma linoleic acid as evening primrose (*Oenothera biennis,* see p. 127). Like that plant, it is used to regulate hormones and relieve pre-menstrual syndrome (PMS).

In Frankfurt, Germany, borage is used to make green sauce (*Grüne Soße*). In Liguria, Italy, it flavors the filling of traditional ravioli and pansoti pastas. The Polish use borage to flavor pickled gherkins.

Greater / Lesser Mountain Mint

Calamints belong to the Lamiaceae family and are found in grasslands and scrub in Europe and central Asia. The genus name comes from the ancient Greek word *kalaminthe*, which means "beautiful mint." This legendary plant is mentioned in a poem attributed to Orpheus, in which it is quoted as having once been a tree, but due to offending Mother Earth it was shrunk to its present form as a punishment.

According to Irish herbal tradition, it was used in the eighteenth century for contraception and to expel a dead child from the womb. It was thought to have abortive properties and contains similar compounds to pennyroyal. Calamints were commonly used in medieval times, but are little used by modern herbalists. Harvest the green mintlike leaves in July for drying and storing. When crushed, they emit a camphorlike odor.

- Both are perennial hardy down to 23°F (-5°C).
- Greater mountain mint is a beautiful garden plant with rosy pink tubular flowers.
- Lesser mountain mint has much smaller white flowers.
- Flowers midsummer through early fall.
- Prefers well-drained sunny position with low nutrient content.
- Cut back after flowering to encourage new growth and second flowering. Grow from seed, cuttings, and division.

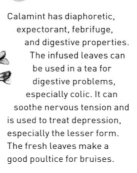

Calamint has diaphoretic, expectorant, febrifuge, and digestive properties. The infused leaves can be used in a tea for digestive problems, especially colic. It can soothe nervous tension and is used to treat depression, especially the lesser form. The fresh leaves make a good poultice for bruises.

Not to be eaten during pregnancy because if taken in excess it can cause a miscarriage.

Lesser mountain mint (*Calamintha nepeta*) is much-prized in Tuscany (where it is known as *nepitella* and *mentuccia*) for its oregano-shaped leaves and flavor combination of mint with a hint of thyme. In Tuscan cuisine it complements porcini mushrooms, rabbit, and the famous ribollita soup. It can also be used to flavor game and wild boar. Add sprigs to olives and to the boiling water for fresh artichokes. It is also excellent for seasoning zucchini, eggplant, and tomatoes cooked with olive oil.

Calendula officinalis

Pot or Garden Marigold

Also known as marybud and marygold, marigold was used by early Indian and Arabic cultures, and also in ancient Greece and Rome, as a protective talisman, medicine, food, dye, and ingredient for cosmetics. The Latin name *Calendula* comes from *calende*, meaning the first day of the month. The names marigold and marybud both refer to the Virgin Mary. The French name, *soucie,* means one who "follows the sun." Some people claim to be able to forecast the weather by observing the behavior of marigolds; the flowers close when wet weather is coming.

In India, the bright sunny flowers are used to decorate the altars of Hindu temples. The ancient Egyptians favored the marigold for its rejuvenating properties, and the Greeks used it as a food decoration. The marigold has been grown in kitchen gardens for centuries and is now cultivated all over the world, although more as an ornamental plant than as a plant for eating.

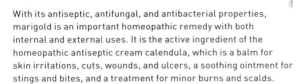

With its antiseptic, antifungal, and antibacterial properties, marigold is an important homeopathic remedy with both internal and external uses. It is the active ingredient of the homeopathic antiseptic cream calendula, which is a balm for skin irritations, cuts, wounds, and ulcers, a soothing ointment for stings and bites, and a treatment for minor burns and scalds.

 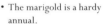

The yellow to deep orange flowers are used in toners for oily skin and are excellent for making a nourishing moisturizer. Marigold shampoo lightens hair and brightens red hair.

Marigold petals are used in salads and as a garnish. They have a sweet nuttiness, while the young leaves have a fresh peppery tanginess. Petals may be dried for use in winter, when they are added as a substitute for saffron in rice, soups, sauces, light sponge cakes, and muffins. The infused leaves also give color to cheese.

- The marigold is a hardy annual.
- Easily grown and propagated from seed, it can be sown every two to three weeks for longer flowering.
- A low-maintenance herb, it is a true sun worshipper that prefers free-draining soil.
- Ideal as an attractive potted or container plant.
- Regular deadheading will encourage flowering all season long.

Chili Pepper

The chili pepper's origins are in Mexico and in Central America. Chili seeds, found at archaeological sites in Tehuacán dating to around 7,000 BCE, are thought to have come from fruits gathered in the wild. The earliest record of actual cultivation came 2,000 years later. The first evidence of chilies being grown in Europe dates to about 1500 CE from where they soon traveled to Asia, India, and Malaysia, then on to China.

Columbus was credited with the introduction of the chili to Europe in 1493. However, it appears that his physician on the second voyage to the West Indies, Diego Alvarez Chanca, was the one who brought the first chili to Spain. He wrote about their medicinal worth the year after. The Portuguese introduced the plant to India at the Portuguese port of Goa.

In his doctrine of signatures, English herbalist Nicholas Culpeper cites chili as under the dominion of Mars—"fiery, sharp, biting taste and of temperature hot and dry to the end of the fourth degree." He suggested using them to treat gas and carbuncles among other ailments. Chilies release endorphins (natural painkillers) in the brain. They contain minerals, including potassium, iron, magnesium, and vitamin B complex. In 1912, Wilbur Scoville developed a means to measure the heat of chilies, which has now been taken over by computers. At the bottom of the scale is the mild bell pepper, while the Dorset Naga is one of the hottest in the world, measuring 1.6 million Scoville heat units. The Carolina Reaper has since taken the lead, registering at 2.2 million.

- Many varieties of chilies are available, hailing from all parts of the world where the temperature is conducive to their growth.
- Chilies need warmth and good light levels to germinate and grow. They dislike cold damp conditions.
- Sow seed in early spring under glass, keeping at a constant temperature to ensure even germination.
- Best sown in individual modules to minimize root disturbance. When roots are fully developed, transfer. As plants grow, keep transferring for plentiful fruits, which appear after the flowers fade.

It is a well-known fact that people who often eat chilies develop a tolerance of their heat, enabling them to eat hotter ones with impunity. Chilies vary greatly in terms of heat. The heat is not related to physical size; some of the smallest chilies, such as the bird's-eye variety, pack the biggest punch. Chilies feature in countless dishes around the world. They can be used fresh, dried, pickled, and powdered. Vegetables, meat, and fish are all enhanced by a chili or two. Chilies are also used in pickles, chutneys, jellies, and sauces. They are even added to chocolate-based products and drinks, a practice common in Mexico.

Chilies are antibacterial and analgesic. They feature in capsicum ointment for use where heat is needed, such as aching muscles and painful joints. They can relieve lumbago and neuralgia. Chilies clear mucus and aid nasal and lung congestion. They are also known to lower cholesterol.

Chili makes an effective insect repellent against ants, caterpillars, and other food-crop pests. To make a fiery spray, grind up three or four chilies with a tablespoon of laundry soap flakes; add to 1¾ pints (1 liter) of water, infuse for thirty minutes, then shake well, decant into a clean sprayer, and use immediately.

In Punjab, some people put pepper in the eyes of the corpse to prevent the ghost from seeing her way back to the house.

SIR JAMES GEORGE FRAZER,
PSYCHE'S TASK (1913)

Caraway

Caraway is native to Europe, Asia, Siberia, Turkey, India, and North Africa. In modern times it has been commercially cultivated in Germany and Holland. The seeds have been found in the remains of prehistoric meals, Egyptian tombs, and along the route of the Silk Road. The common name is thought to derive from the Arabic *al-karwiya*, which is probably where the Latin word *carro* comes from, meaning cart or wagon.

Caraway has been used continually by humans for food and medicine for thousands of years so that it has become something of a revered herb, with many associated superstitions. In German folklore a dish of caraway was placed under a child's crib to ward off witches. It was a common ingredient in love potions designed to prevent lovers from being fickle. Another belief held that baking the seeds in a cake or bread would inspire lust; caraway seed cake was once popular at weddings.

Sweetly spiced caraway seeds were chewed or infused as a tea to aid digestion and stimulate the appetite, as well as to sweeten the breath. They were traditionally served after Elizabethan feasts. Herbalist John Gerard wrote "caraway consumeth wind and is delightful to the stomache." Today caraway is still recommended for flatulence and in some cultures the seeds are offered after the meal as we would serve after-dinner mints. Caraway was also once recommended as a herb to promote mental powers and the memory.

- A hardy biennial in the family Apiaceae, which includes carrot, celery, and parsley, it is similar in appearance to other members of carrot family. Grows feathery leaves in the first year, flowers in the second year, and produces umbels of white and pink flower heads.
- Sow fresh seeds outdoors in early fall in rows or blocks.
- Prefers full sun but can take a little shade if the soil is well-drained.
- Sun aids strength of flavor in the ripe seeds.

Use young leaves fresh in salads. Roots can be cooked as a vegetable, by steaming or roasting. The seeds are commonly used in breads and in caraway seed cake. They are an ingredient that gives sauerkraut its distinctive flavor, and the German liqueur kümmel contains oil of caraway together with cumin. Caraway features in Hungarian beef stews. Rich meats, such as pork and goose, benefit from its flavor and digestive qualities.

Caraway is rich in antioxidants and has numerous health benefits. It has analgesic, antispasmodic, carminative, and digestive properties. It can be taken as an infusion to relieve colds, congestion, and bronchitis. It is also calming for the digestion and can be used to treat gas and bloating. The seeds contain a potent antiseptic agent that prevents and cures infection. Oil of caraway is used to treat rheumatism, aching joints, and toothache.

Peas will benefit both in quality and taste with *Carum carvi* planted as their neighbor in the same row. Never plant caraway with fennel because each will inhibit the growth and health of the other.

Do not take caraway in high doses if you are suffering with any kidney or liver problems because of the content of the volatile oils.

Cedronella canariensis syn. *C. triphylla*

Balm of Gilead

A number of plants have shared the name "balm of Gilead." *Commiphora opobalsamum* was famously the tree that the Queen of Sheba gifted to Solomon. The plant is referred to in the Bible: "Is there no balm in Gilead; is there no physician there?." Today it is very rare, protected, and illegal to export. The tree *Populus balsamifera* is also commonly known as balm of Gilead.

The herb *Cedronella canariensis*, as its name suggests, originates from Maderia, the Azores, and the Canary islands. It is also found growing in the wild in North Africa and has become naturalized in several places, including South Africa, New Zealand, and California.

C. canariensis is a perennial, flowering herbaceous plant. It is a member of the Lamiaceae family and has a very distinct balsamic, camphorlike scent; when the leaves are crushed they immediately release head-clearing, resinous vapors. This form is also known as "false" balm of Gilead, canary balm, and Canary Island tea.

- Balm of Gilead is a half-hardy, perennial, semi-evergreen plant, with three lobed, toothed, pointed mid-green leaves that exude a musky lemon camphor scent with a hint of eucalyptus. It produces pink-pale mauve, two-lipped flowers.
- Can be grown from seeds and cuttings.
- Can be grown outside from late spring to early fall. Not frost hardy; it needs protection through the winter.
- Grows happily in a container. Keep transferring as it grows.

Cedronella has medicinal properties: antidepressant, aromatic, and decongestive. Use an infusion of the leaves and flowers to treat coughs and colds. Crush the leaves and inhale to clear sinuses and muzzy heads, and revive flagging spirits. Leaves can be rubbed on the skin to deter mosquitoes, ants, and other biting insects. It has also been used to treat aching joints, bruises, and sprains.

For a soothing, uplifting, and invigorating bath, add a tied bunch of fresh leaves to infuse in the hot water. The leaves will yield aromas of lemon, balsam, and eucalyptus.

Dried leaves and flowers of balm of Gilead make an interesting addition to a woody winter potpourri, mixed with cloves, cinnamon sticks, and dried juniper berries.

Centaurea cyanus

Cornflower

The cornflower is well known today as a cottage garden and cutting flower with its starlike flowers of brilliant azure blue. It was the favorite flower of the goddess Flora (Cyanus) and its genus name is derived from the centaur, Chiron, who taught healing using herbs.

Whereas Britain has the red field poppy, France has the blue cornflower as its symbol for Armistice Day (November 11). Cornflowers were as endemic as poppies in the cornfields of Flanders. Modern agricultural methods have more or less eradicated its presence from even the edges of cornfields. The blue cornflower has also been a national symbol of Germany and Estonia.

The alternative name of Bachelor's Button comes from the practice of young men wearing the flowers in their lapels to advertise their availability. Flowers of the cornflower were once used for dyeing, making a true blue ink and watercolor. The dried flowers have always been a colorful addition to potpourri.

The edible flowers are used in salads and to decorate food. Also used in tea blends, most famously Lady Grey.

Cornflower is astringent, anti-inflammatory, and sedative. Use in an eye wash for corneal ulcers, conjunctivitis, and sore eyelids. As an astringent herb, it helps reduce inflammation and can be used to treat minor wounds and mouth ulcers.

The famous French eyewash *Eau de Casselunettes* was made from the flowers of the cornflower because of their ability to "brighten the eyes." Facial steams and bath preparations have also benefited from the use of the flower petals. Used as a hair rinse for gray hair, it adds luster and brightens.

A member of the Asteraceae family, the almost flat daisy-like flowers, which appear from June until August, provide a convenient landing stage for any visiting bee, offering easy access to nectar and pollen-rich flowers.

- An easy to grow hardy annual.
- Sow seed in fall or spring either in situ in the border or into individual modules for planting outside when roots are well established.
- Likes a well-drained, sunny position.
- Several cultivars have varying pastel shades, including pink, purple, and lavender.

Chamomile

Chamomile (also spelled camomile) comes from the Greek words *khami* ("on the ground") and *melo* ("apple or melon") due to its growing habit. It was one of the most popular strewing herbs due to its sweet apple scent. The common name "Roman chamomile" was bestowed on the plant by the sixteenth-century German scholar Joachim Camerarins, who observed it growing profusely near Rome. Chamomile was so admired by the ancient Egyptians as a "cure-all" that they offered it as a dedication to their sun god, Ra.

The herb has been naturalized in England since the Dark Ages, when it was known as "Maythen." It was recorded in the eleventh-century Anglo-Saxon ancient manuscript *Lacnunga*, a collection of medical texts and remedies, and was one of the nine sacred herbs of the early Saxons. *Chamaemelum nobile* is known as the true chamomile of the British Isles.

One for the herbal pillow, mixed with lavender and hops, to soothe, relax, and promote sound sleep. Stuffed into a pet's bedding, it keeps fleas at bay.

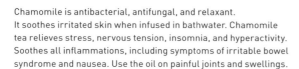

Chamomile is antibacterial, antifungal, and relaxant. It soothes irritated skin when infused in bathwater. Chamomile tea relieves stress, nervous tension, insomnia, and hyperactivity. Soothes all inflammations, including symptoms of irritable bowel syndrome and nausea. Use the oil on painful joints and swellings.

Chamomile has long been used as a final rinse for those with fair hair to brighten and add luster. It has also been a popular herb used by the cosmetic industry, especially oil of chamomile, in soaps, shampoos, and bath products.

Chamomile is known as a "plant doctor"—it increases the growth and vitality of neighboring plants. Cabbages and onions thrive with chamomile as their companion. It can be used to make a delicious syrup to feed to bees.

- Hardy, perennial evergreen herb.
- White flowers with yellow centers appear all summer.
- Sow seeds in the spring in situ or in modules.
- Divide in the second and third year in spring to prevent weakness.
- Cut back after flowering and feed; due to their high aroma, they are pest free.

Chenopodium bonus-henricus

Good King Henry

Good King Henry is a perennial plant that is native to Europe, western Asia, and North America. It is found growing wild on wasteland and roadsides and is cultivated as a vegetable and a medicinal plant. Although naturalized in North America, it is not commonly used there. It spread throughout Europe thanks to the Romans, who introduced it to their conquered territories.

The generic name comes from the Greek word "chen," meaning goose and "podus," meaning foot, referring to the shape of the leaves, which are webbed like goose feet. Its specific name refers not to the English king Henry VIII, but to the French king, Henry IV of Navarre. Good King Henry could, however, be found growing as a crop in Tudor gardens.

For many centuries the herb has for been known as a subsistence food, often gathered from the wild by poorer country folk in need of nourishment. At one time it was grown extensively in Lincolnshire, England, where it was also known as Lincolnshire spinach or asparagus. It is a plant that is favored in a permaculture system.

- Seeds can be erratic—they take time to germinate.
- Sow seeds directly into prepared soil in fall or early spring.
- Once growing, it needs little or no attention.
- Cut back after seeding, bury the compost, feed, and wait for next crop of leaves.
- Can be divided in spring to stop the heart of the plant becoming woody.
- Allow new young plants at least one year to establish before harvesting.

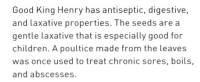

A good herb for pots. Use young leaves in salads. More mature leaves can be steamed in the same way as spinach. A great "pick-me-up" herb, it contains high levels of iron, calcium, vitamin B1, and vitamin C. The whole plant above the ground can be eaten. The flowering tops can be cut, steamed, buttered, and eaten as you would asparagus. Young stems can also be steamed and used for dipping into the Provençal garlic sauce, aioli.

Good King Henry has antiseptic, digestive, and laxative properties. The seeds are a gentle laxative that is especially good for children. A poultice made from the leaves was once used to treat chronic sores, boils, and abscesses.

Cichorium intybus

Chicory

Common chicory, also known as succory and blue endive, is native to Europe, but has been introduced to many other countries, such as the United States where it has naturalized. It is found growing wild on roadsides, field edges, and wastelands. Cultivated for its root, leaves, and flowers, it is has been used as a herb since ancient times.

Egyptians used the leaves and flowers in salads and prepared the roots as a vegetable. According to Pliny and Theophrastus, it was also relished by the Greeks and Romans. The French have cultivated it for centuries; the long taproot is dried, ground, and mixed with coffee to counteract the stimulating effects of caffeine. It has also long been used as a substitute for coffee.

The Swedish botanist Linnaeus used chicory as one of the flowers of his famous floral clock at Uppsala because the flowers open when the sun rises and close when the sun is at its highest around noon. A familiar plant of folklore, it was perceived to have magic powers; the flowers were steeped in water, which was then used to see into the future as well as to improve eyesight.

Chicory leaves have been boiled to produce a blue dye. It has also been grown as a crop for animal fodder. Brewers have added roasted chicory to stout to give a coffee-like flavor

Wild chicory leaves are usually bitter. Use young leaves, flower buds, and petals in salads. The dried . ground root is used as a substitute for coffee. Roots can also be eaten fresh, boiled, and served with a sauce as a side vegetable.

Diuretic and laxative, chicory has a general cleaning action. The root in particular is used to treat bladder problems. As bitterness increases the flow of bile, it has been used as a tonic, specifically for gallstones and the liver. Its diuretic properties make it a good herb for rheumatism and gout, as it eliminates uric acid from the body.

Excessive use of chicory may impair the functioning of the retina of the eye.

- Hardy perennial.
- Sow seeds in the spring or fresh from the plant for quick germination straight into the soil or seed tray. Plant outside when roots are fully developed.
- Dig roots in the fall for root cuttings and for harvesting for drying.
- Flowers July to October.
- Loved by bees, butterflies, and hoverflies.

Convallaria majalis

Lily of the Valley

This native plant of Europe, Asia, and North America is occasionally found growing wild in shady woodlands and clearings; in gardens it is prized for its dainty, highly scented, white bell flowers, making it a favorite ground-cover herb for shady spots. The name *Convallaria* comes from the Latin *convallium*, meaning "of the valleys," while *majalis* comes from *maius*, the month of May, when it flowers. It was formerly known to apothecaries as *Lilium convallium*.

In the Middle Ages, the herb came to be used in bridal bouquets, the flower symbolizing modesty and purity. It later became a favorite with the Elizabethans for use in nosegays, "tussie mussies," and posies and was carried as a means of enduring the often noxious air.

In the language of flowers, lily of the valley means "the return of happiness." The flower was often depicted on the ornate postcards that servicemen and their sweethearts exchanged in World War I. The plant had another benefit for servicemen in that war: its roots were used to make an ointment to treat burns inflicted by mustard gas.

At the Gaelic Beltane or May Day festival, young women would gather lily of the valley flowers and other spring blooms into baskets (lined with moss to keep the flowers fresh) and deliver them to sweethearts as a love token. Flowers were also used to decorate maypoles.

The delicate scent of lily of the valley has inspired perfumiers for centuries. Floris, Coty, and Yardley are among the many famous houses that have sold a perfume based on the herb.

The herb contains cardioactive glycosides that strengthen the heartbeat and slow it when it is too fast. It is also used for urinary obstructions and as a diuretic. Lily of the valley is still an important drug in some pharmacopoeias.

All parts of the plant are poisonous and should only be prescribed by a qualified herbalist. Lily of the valley is subject to legal restrictions in some countries.

- This hardy perennial is best grown from divisions of the crowns in the fall.
- The plant prefers woodland shade; in the garden, choose a shady area with few competing plants as it can be difficult to establish.
- Prefers humus-rich soil, where it will eventually spread indefinitely.
- Poorly flowering plants are usually overly congested or overfed; reduce the amount of manure they receive.

Coriandrum sativum

Coriander, Cilantro

Coriander is one of the most ancient herbs and has been cultivated for at least 3,000 years. References to it can be found in Sanskrit texts, Egyptian papyri, and in the Bible, in which "manna" is compared to coriander seeds. Egyptian mourners placed jars of coriander in Tutankhamen's tomb to assist his spirit on its journey to the Land of the Dead.

The herb is native to regions from southern Europe and North Africa to southwestern Asia. Coriander seeds were taken from the Mediterranean in the caravanseria along the Silk Road to China, where it enriched the cuisine and its profile was raised from a mere flavoring to having immortal powers. It appears to have reached the shores of Britain in the late Bronze Age and evidence of seeds has been found in the remains of ancient huts.

Coriander (the leaves are often known as cilantro) was one of the herbs the early North American settlers planted, grew, and used, as recorded in 1670 in Massachusetts. In the sixteenth century the Spanish conquistadors introduced it to Mexico and Peru, where it has become a much-loved culinary partner to the chili pepper.

- Grow plants for leaf and seed. Prefers light, well-drained soil in sunny position and dry conditions.
- Grow from seed in spring in drills covered with a light sprinkling of soil. Germination usually takes place within five to ten days.
- Can be started off in modules, minimizing the root disturbance that causes it to bolt to seed quickly.
- Sow seed from spring to early fall at intervals for continuous crops.

Coriander is used in countless cuisines all around the world, including Indian, Chinese, Mexican, African, and European. The fresh leaf and the dried seeds are most commonly used. The dried and ground seeds are an ingredient of garam masala, the mixed spice base used in many curries; also used in tagines, soups, chutneys, hot pickles, and preserves. It is also sometimes added to cakes, cookies, and candies. Fresh cilantro leaves are usually added to cooked dishes just before serving because flavor is quickly lost. The roots are used in some Thai dishes.

When planting, keep well away from fennel, which inhibits the growth of coriander, which has been known to wilt and wither in its presence. It is happy planted next to dill, chervil, and anise and improves their growth.

A strong infusion of the whole coriander plant can be used as a spray to counteract the march of the red spider mite, a common sap-feeding pest, especially on glasshouse plants.

Coriander has a number of medicinal properties: antibacterial, anti-inflammatory, digestive, expectorant, and fungicidal. Chew the leaves to alleviate nausea and the seeds for mild digestive complaints. The seeds have a faintly orange scent and can also be chewed as a breath freshener. Make an eye bath to reduce swelling, pain, and conjunctivitis. Oil of coriander is used as a massage oil for muscular aches and pains. Use in a poultice for painful joints.

> *If I had a palace made of pearls, inlaid*
> *with jewels, scented with musk, saffron,*
> *and sandalwood . . . Your Name*
> *would not enter into my mind.*

SRI GURU GRANTH SAHIB, CENTRAL TEXT OF SIKHISM

Saffron

The generic name *Crocus* is from the Greek *krokos*, meaning a thread, a reference to the plant's threadlike pistils. The common name of saffron comes from the Arabic *zafaran*, meaning yellow. Saffron's beauty and color are often mentioned in Greek mythology and poetry, and both the color and aroma were greatly admired.

Alexander the Great, while on his campaign in Asia, was known to have coveted saffron and used it in infusions; he also added it to his bath water to heal battle wounds. The practice was taken up by his troops, who introduced saffron bathing as a pleasure on their return to Greece. The Egyptian queen Cleopatra was supposed to have bathed in saffron water as an aphrodisiac.

The Assyrians, Sumerians, Egyptians, and Minoans prized this herb above all others. To this day, it is the most expensive spice in the world, because it is still cultivated and harvested by hand, usually by the women of villages, as it was in ancient times. Iran is the now the biggest producer, but it has been grown in Europe since the medieval period. In England it was grown around Saffron Walden in Essex, where is was mainly used for dyeing. Saffron from Valencia, Spain, is now considered to be the best in the world for quality, flavor, and color.

- It takes three years to grow a productive corm from seed.
- It is better to use mature corms that can be divided. The corm offsets should be separated in late spring.
- Saffron grows well in rich, sandy, well-drained soil in a sheltered position, in full sun.
- The plant thrives but fails to flower in areas and years with poor summers.
- Harvesting of the pistils takes place when the flowers are open and the pistils are easily accessible.

Saffron was highly prized as a dye, used to color the robes of Buddhist monks, because yellow was the color associated with sacrifice, quest for light, and salvation. Independently, the dye was used by the Persian kings, whose saffron-dyed shoes were part of their regal attire.

Saffron tea is a warming, soothing drink that eases the digestive tract and induces perspiration; it is used to treat coughs and colds. Saffron is also thought to heal and ease an ailing liver, and it was once given to ease melancholy and hysteria. In the Italian renaissance, Venetian women dyed their hair blonde with a mixture of saffron and lemon juice that was baked in the sun.

Saffron is expensive to buy as a culinary spice, but only a tiny amount is needed to color a dish and flavor it with its pungent earthy essence. Paella, risotto Milanese, many Indian rice dishes, and Cornish saffron cake and buns all include it. Cooks should check that their saffron has not been adulterated with turmeric or safflower, which bear no comparison to the real thing.

Producing good saffron is a race against time. Each flower has three stigmas, and between 50,000 and 75,000 flowers yield just 1 pound (450g) of dried saffron. Once picked, saffron must be dried very quickly because it spoils within minutes. Only 50 tons are produced each year.

Cryptotaenia japonica

Japanese Parsley

Japanese parsley, also known as Japanese honeywort and white chervil, is thought to have originated in Japan, from where it was carried to China, Taiwan, and beyond. The plant is a member of the Umbelliferae family, which includes other forms of parsley, dill, and coriander. It is now cultivated as a culinary herb in Southeast Asia and occasionally in North America. The herb has the names of *mitsuba* or *mashimori* in Japanese and *san ye gin* in Chinese, both of which refer to the plant having three leaves. One other name in China is *ya er gin,* meaning "duck celery."

The plant has an attractive form with its pretty, flat, three-leaved leaflets, which are light or dark green, and slightly heart-shaped leaves with sawtoothed edges. Irregularly spaced on the parent stems are flower stems topped by tiny white flowers. The genus name *Cryptotaenia* comes from the Greek *cryptos,* meaning "hidden," and *taenia,* meaning "ribbon"—the name alludes to the appearance of the fruits.

- Sow seed in late spring, but be patient as it can be slow to germinate.
- A woodland plant by origin, it prefers a shady position in moisture-retentive soil. Soil that is humus-rich gives good leaf production.
- The perennial is hardy down to 14°F (-10°C); commercially, it is grown as an annual cropping salad herb.
- It freely self-seeds and makes good edible ground cover.
- Cut flowers back and feed to encourage more leaves.

This classic seasoning herb is popular in all Asian cuisines. All parts are edible, even the roots. Resembling flat-leaf parsley and with a flavor suggesting angelica, chervil, and celery leaf, it is used both as an herb and salad green. It is an ingredient of miso soup, salads, sushi, sukiyaki, and sashimi, and is mixed into tempura batter, savory and sweet custards, and rice dishes. Japanese chefs use the tender young stalks to tie up sushi rolls.

Japanese parsley is anti-inflammatory and was once used for the treatment of colds, coughs, and fevers, and as a tonic for strengthening the body after a period of illness.

Japanese parsley is widely grown in herb gardens, but it is the purple-leaved form, *C. japonica* f. *atropurpurea,* that is most often used as an ornamental plant.

Cumin

A native plant of Egypt, cumin has long been valued for its unmistakable aroma and flavor. Preserved seeds have been found in the tombs of the pharaohs, and cumin was used in mummification. From Egypt cumin traveled to Greece and the rest of Europe, and to Asia. Taken to the Americas by the Spanish and Portuguese, it is especially highly valued in Mexico.

Cumin is an essential spice in Indian cuisine, appearing consistently in spicy aromatic dishes and curries that otherwise vary from region to region. In North Africa, notably Morocco, its taste is essential to tagines and other lamb dishes. India and Morocco are large-scale cumin exporters, and the spice is also grown on the Mediterranean islands of Sicily and Malta.

At one time, cumin was sufficiently highly valued to be carried in the pocket and used for bartering. The seed came to be associated with fidelity, and was carried at weddings as a sign of commitment. Equally, young women would give their beaux bread or wine seasoned with cumin to keep them faithful.

- Cumin is grown from seed sown in early spring. A temperature of above 60°F (16°C) is required for germination to take place, usually within five to ten days.
- Sow in modules or small pots to minimize root disturbance.
- Plant outside only when the temperature reaches 60°F (16°C). This half-hardy annual will struggle in cold and damp conditions.
- Plant in well-drained, fertile soil.
- The seeds are ready for harvesting when they turn yellow-brown in color.

Cumin is an ingredient of garam masala and other mixed-spice curry powders. In Southeast Asia it is used in curries and pickles, and roasted as a "sprinkling garnish." It is used in North African dishes, such as lamb, couscous, yogurt, and eggplant. With chopped mint, cumin flavors salads of cucumber and carrot.

Since Biblical times, cumin has been added to wine, water, and bread to soothe and calm the digestion. It is used in Ayurvedic medicine to improve liver function, and also has a carminative role in veterinary medicine.

Cumin seed steeped in wine makes a lust potion.

"CUMIN" ON EARTHWITCHERY WEBSITE

Lemongrass

Although there are many *Cymbopogon* species of half-hardy perennial grasses, the best known is *C. citratus*, sometimes called West Indian lemongrass, which also goes by the name of citronella, due to its ability to repel insects. A native of Indonesia, this tall tropical grass grows in large clumps and exudes a powerful lemon fragrance. Today, it is grown in many parts of South America, Africa, and Indochina. Another species, *C. flexuosus*, known as East Indian lemongrass or cochin, is grown for its production of high-quality oil, which is used in perfumery.

The Brazilian Quilombolas tribe traditionally used lemongrass as a means to calm the over-excited. The Carib of Guatemala recognized its carminative properties and used the leaves in a tea to relieve flatulence and gripe. It was once used to flavor tobacco. Its use in modern-day commodities are many, with the oil appearing in cosmetics, soaps and bath preparations, detergents and other household products, foodstuffs, confectionery, and modern medicine.

Lemongrass also has a role in environmental protection. It is widely planted on embankments or dikes in South and Southeast Asia, as a means of soil conservation. The leaves are widely used as a mulch.

- It can be grown from seed, but keep the temperature above 68°F (20°C) for successful germination.
- Bought stems can be placed in water until roots appear, then transferred to grow on.
- Needs a well-drained position in full sun.
- Grows well in pots filled with a good-quality humus-rich compost.
- In cultivation, it flowers only in the tropics.
- Plants need to mature before the stems can be used.

Lemongrass has antioxidant, anti-inflammatory, antifungal, and antiseptic properties. It is traditionally used in the form of a soothing and calming tea to treat digestive problems, especially stomachache, cramps, and vomiting. Lemongrass oil is also applied externally to ease joint pain, aching muscles, and neuralgia. The herb can also improve blood circulation.

A natural insect repellent, lemongrass can offer some protection from mosquitoes, especially when the essential oil is used in an oil burner or mixed into the wax of a candle. Lemongrass can protect pets too; make a water spray with a few drops of lemongrass oil and eucalyptus, then spray the pets' bedding to keep it fresh and free of fleas.

Lemongrass is much prized in numerous Southeast Asian cuisines, especially in Thailand, Vietnam, and Indonesia, for its distinctive sweet-sour lemon flavor. Leaves may be picked throughout the growing season, and whole stalks or small bunches added to infuse in a dish, then removed before serving. Stalks may be bruised to release their fragrance and added to dishes with a base of coconut milk. In Bali, Indonesia, the stalks are used as skewers to barbecue seafoods.

Lemongrass should not be consumed during pregnancy because it seems able to trigger menstrual flow, and thus could cause miscarriage. It is also not recommended for women who are breast-feeding.

*Our pasta tonight is a squid ravioli
in a lemongrass broth. God, I hate
this place. It's a chick's restaurant.*

BRET EASTON ELLIS,
AMERICAN PSYCHO (1991)

Dianthus caryophyllus

Clove Pink

The clove pink, also known as the clove carnation, gilly flower, or sops in wine, is a native of central and southern Europe that has been cultivated for its scent for more than two thousand years.

The genus name *Dianthus* comes from the words *dios*, meaning divine, and *athos*, meaning flower. The plant was given its name by Theophrastus (370–285 BCE), a botanist living in ancient Greece. Both the Greeks and Romans held the plant in high esteem, making coronets and garlands from the flowers. Accordingly, the name "carnation" has the same origin as "coronation."

The clove pink made its first appearance in Britain in Norman times, and was eventually carried to the New World by the European colonists. It was once used in tonic cordials to treat fevers, and the common name "sops in wine" derives from this usage. Another common name, "tussie mussies," is a reference to the strength of its curative clove scent. Today, the clove pink is used in the making of some of the most expensive perfumes in the world by Floris, Nina Ricci, Estée Lauder, and others.

In earlier centuries, clove pink petals were used instead of cloves, which were much more expensive because they had to be imported from the Spice Islands of Indonesia. The petals may be added fresh to salads, crystallized for use in cake decoration, or made into a cordial. If using the petals fresh, remove the white head because it is relatively bitter.

The English herbalist John Gerard recommended clove pink for flatulence and heartburn, saying, "a conserve made of the flowers with sugar is exceeding cordial, and wonderfully above measure doth comfort the heart, being eaten now and then."

Dried petals are added to potpourri, scented sachets, almond oil, and wine vinegar. The petals should be macerated and left for several hours to achieve a strong scent.

- Grow from seed or cuttings from non-flowering shoots in summer.
- Clove pink is a herbaceous perennial that requires a well-drained position in full sun to thrive.
- Prefers neutral to alkaline soils.
- A good plant to grow massed in pots, especially when sited in areas where their heady perfume can be appreciated.
- Harvested flowers will be replaced in healthy plants. Dead-head regularly to encourage good production of fresh flowers.

Purple Coneflower

The purple coneflower, also known as Sampson root, echinacea, and red sunflower, is a herbaceous perennial native to eastern North America; it is now found growing wild in the eastern, southeastern, and midwestern United States. An earlier name for it was *Rudbeckia purpurea*. The current generic name *Echinacea* comes from *echinos*, the Greek for hedgehog, a reference to the spiky center of the flower at the seed stage. Echinacea was originally used by Native American tribes, such as the Comanche, Choctaw, and Delaware, to treat wounds. It is now considered one of the most effective detoxicant herbs in Western medicine. Some tribes also used it to treat insect bites, stings, and snakebite, and they would chew the root to relieve the pain of toothache. It was even used to treat venereal disease. Early settlers adopted the Native Americans' uses of the plant, and it remained in folk medicine until it was patented by Dr. H. C. F. Meyers, who claimed it had blood-purifying properties and was therefore useful in treating the bites of rattlesnakes and other American snakes.

A relative, *E. angustifolia*, is reported to have stronger constituents and is therefore more favored in America for its healing properties.

- All *Echinacea* species can be grown from seed.
- A mature plant may be divided in spring or fall.
- The plant is easiest to grow in cool, damp climates.
- It prefers a well-drained position in full sun, and is drought resistant once established.
- The roots of mature plants are dug up in the fall and are cleaned and dried before being used medicinally as a tincture, infusion, tablet, or powder.

With its flowers appearing from high summer in a wide range of shades, purple coneflower has become a popular border plant. It offers good potential for breeders and is attractive to bees and butterflies.

Echinacea tincture should not be taken continually as a cold preventative because it may have an adverse effect. In particular, it can interfere with the action of some medications.

Medicinally, echinacea is antibacterial, antifungal, anti-inflammatory, and antiviral. It is used to remove toxins from the circulatory, lymphatic, and respiratory systems, and as a gargle for sore throats and tonsillitis. A salve made from the root may be used to treat acne, psoriasis, minor cuts, and infected wounds. Tincture of echinacea has been shown to stimulate the immune system and is best taken at the onset of a cold or viral infection to reduce its impact.

Echium vulgare ⚠

Viper's Bugloss

An exotic-looking plant that originated in the Mediterranean, viper's bugloss spread rapidly throughout Europe to North America, seeding easily in wastelands and scrub. It has since become something of a pest to farmers in the United States. It was used by Native American tribes such as the Cherokee and Iroquois for urinary complaints.

Also known as blueweed, wild borage, and snake flower, it gained its name from its ability to expel poisons and venom. The plant may be said to conform to the Doctrine of Signatures—its seeds are said to resemble the head of a viper, and the flowers to have the appearance of a snake's head, so it follows that it should be used to treat snake bites.

The common name "bugloss" originates in the Greek for ox's tongue and alludes to the roughness and shape of the leaves. The flowers produce nectar for months on end, making the plant highly attractive to bees, butterflies, and other insects.

Eating the leaves is not recommended due to their many fine hairs, which can be an irritant not only to the throat but also the skin.

Viper's bugloss flowers are eaten in salads, or crystallized and made into cordials and flower drinks. Eating the rest of the plant should be avoided because it, and a related plant, *Echium plantagineum*, are associated with livestock poisonings. The leaves can be brewed into a tea.

- This hardy biennial flowers and self-seeds freely in its second year.
- It is better not to compost spent plants because any seed is likely to survive and cause rapid spread. Viper's bugloss is quick to invade.
- An excellent species for wildflower planting, and it also grows in containers.

The roots were once used to produce a red to purple fabric dye. An olive-green dye may also be obtained from the flower stalks.

Viper's bugloss is diuretic, demulcent, and anti-inflammatory in its effect. Fresh leaf tips are used for making poultices to treat boils. Infused young leaves encourage perspiration in those with fever. Mixed with wine, the seeds were once used "to comfort the heart and drive away melancholy."

Cardamom

Cardamom, a native of evergreen tropical forests of India, Nepal, and Bhutan, is an evergreen, clump-forming perennial and a member of the ginger family. For thousands of years it was one of the most sought-after edible commodities and was carried overland by camel caravans along the Spice Road, which connected India with China and Europe. It was popular not only as a culinary spice but also as a component of perfumes.

The plant was introduced to Southeast Asia more than a thousand years ago, and is now found growing in ruins of the ancient trading posts of Khmer (Cambodia). It has long been cultivated commercially in India, Sri Lanka, and Guatemala.

A record of 720 CE suggests that it was used in Chinese medicine; in Ayurvedic medicine it was known as "ela" and used for bronchial and digestive problems. But cardamom has never been the preserve of medicine men alone; traditionally, it was used as an aphrodisiac in love potions. And it has improved relations between people in another important way. If chewed, the pods are an extremely potent breath freshener.

- Seed must be fresh for germination to occur.
- Given a minimum temperature of 75°F (24°C), growth is rapid.
- Grown as a potted plant, cardamom can be protected from frost by being moved into a glasshouse or conservatory kept at a minimum temperature of 64°F (18°C).
- Prevent sudden temperature drops. Cardamom needs warmth, good light levels, and humidity to thrive.

Cardamom is an anti-inflammatory and antioxidant agent. Due to its warming properties, it is a good aid to digestion and combats nausea and vomiting. It is also a purifying agent and plays a part in detoxification programs.

Like vanilla pods, cardamom pods are used whole for steeping in liquids to infuse their flavor and aroma for baking purposes, fruit pies, and milk desserts. The pods are slit and the seeds are removed, pounded, and ground for curries and rice dishes. Cardamom is used for traditional Indian desserts and masala chai (tea), and is added to coffee in the Middle East. Scandinavians use the pods in hot toddies and mulled wines. Cardamom is also a popular flavoring in cocktails.

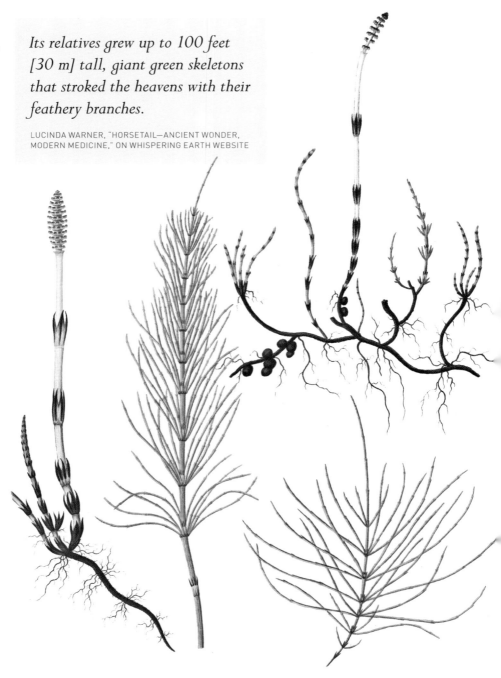

Its relatives grew up to 100 feet [30 m] tall, giant green skeletons that stroked the heavens with their feathery branches.

LUCINDA WARNER, "HORSETAIL—ANCIENT WONDER, MODERN MEDICINE," ON WHISPERING EARTH WEBSITE

Common Horsetail

A herbaceous perennial plant native to Europe, common horsetail is also found growing in the wild in China and North America. Both the generic and common names refer to the form of the plant, the Latin *Equisetum* deriving from *equus* for horse and *seta* meaning bristle.

Common horsetail comes from a large genus of plants with origins in giant varieties that existed in the Carboniferous period (about 360 million years ago) and that are widely preserved as fossils. The horsetail of today is simply a dwarf form of those ancient plants.

The horsetail has a unique ability to absorb silica from the soil. This gives the plant a harsh texture and feel, and from Roman times until the eighteenth century it has been put to use as a natural aid for scouring pots and utensils. Blacksmiths and metal workers of old would polish swords and other weapons with horsetail, trusting that the strength of the plant would transfer into the piece. In particular, the stems of the plant were judged suitable for polishing pewterware, and that usage is recalled by one of the plant's common names, pewterwort.

The discovery that horsetail absorbs gold into its cells has made it a useful diagnostic tool for gold prospectors, who can analyze the ashes of burned horsetail to pinpoint the presence of local gold deposits.

- Horsetail is a cryptogam, a plant without flowers or seeds, that relies on floating spores like that of ferns to proliferate. It also spreads by fast-creeping rhizomes.
- The spores are found in conelike catkins produced early in the year.
- It prefers wild habitats of moist loam or sandy soils, usually near water in fields or hedgebanks, wastelands, and among sand dunes.
- It is best grown in a large container because the roots can be very invasive.

The concentration of minerals in horsetail lends it to treatments of the hair and nails. A decoction may be used in a nail bath for strengthening nails, and the herb makes an excellent hair rinse, tonic, and natural conditioner, one that leaves the hair healthy and shining.

Horsetail can break down hardened deposits in the body and was therefore used in medicine to treat arteriosclerosis, arthritis, cysts, gout, and kidney and urinary stones. By dissolving such deposits it can promote ease of urination, tone the bladder, and relieve water retention. Used in the form of a compress or poultice, it can staunch bleeding.

Common horsetail is antiseptic and anti-inflammatory and is nearly always used in combination with other herbs, especially when used for its antiseptic properties. The stems are gathered in early summer after the fruiting stems have died back. They are then crushed to allow water held in the joints of the plant to drain away, after which they are dried and stored. Horsetail is rich in minerals and is used in infusions for enriching the blood, and its high levels of calcium and silica help to support bones and tissue.

The whole plant can be used to extract a natural dye that gives a light yellow-ochre color with a mordant of chrome or alum. In the garden, horsetail "tea" is used prevent apple scab and powdery mildew fungus.

Tasmanian Blue Gum

Eucalyptus is a genus of some three hundred indigenous trees of Australia, with *E. globulus*, commonly known as the blue gum, being the most widely distributed and favored of them all. It has been introduced successfully to southern Europe, Algeria, Egypt, Tahiti, South Africa, and India. It was planted in California, originally for use on the railways. In Britain it is largely limited to gardens and arboreta.

For most people, the Tasmanian blue gum is synonymous with Australia and in particular the Blue Mountains to the west of Sydney. Not only do the trees make up the blue forest that grows there, but they generate a blue haze that hangs in the air above them, created by droplets of oil from the leaves. The haze is beautiful but also hazardous because the oil is highly volatile and the cause of life-threatening fires.

The Tasmanian blue gum was discovered to have one particular advantage: its roots have a powerful drying effect on the soil in areas of marshland, including those that have long been breeding grounds for the mosquitoes that transmit malaria. The discovery of this useful property has transformed wide areas of countries such as Algeria and Italy, making them safe from malaria and therefore safer to inhabit.

- *Eucalyptus globulus* may be propagated by seed in a heated glasshouse in spring or fall.
- The tree prefers a fertile soil, and tolerates acid and alkaline environments.
- It grows in dry, moist, or wet soil and can tolerate drought.
- In a sunny position the tree can grow to 180 feet (55 m), but it won't tolerate shade.
- The tree survives temperatures as low as 5°F (-15°C).

Dyes can be made from the leaves of all eucalyptus species, with the colors varying according to the mordant used. Steeping the fresh leaves of *E. globulus* can produce a vibrant red dye; boiling the bark yields a beige dye.

The leaves of *E. globulus* contain small translucent glands, and these are the source of eucalyptus oil, which is widely used for its decongestant effect. The oil is used for treating catarrh, bronchitis, and influenza. Traditionally, it brings benefits as a vapor bath, chest rub, or inhalation. The oil is also antiseptic and anti-inflammatory and so used in liniments for bruises, sprains, and muscular pains as well as dressings for wounds and rheumatism. Eucalyptus oil is effective as a massage oil for painful joints. A powerful disinfectant, the oil is used in many cleaning products.

In temperate climates, *E. globulus* is the most-used eucalyptus variety that is pulped for paper making. The tree is also widely used for fuel, while the timber has numerous applications, including flooring, furniture, and boat building. The wood is also used for pit props in mines. In Australia, the indigenous aborigines had a less prosaic use for the tree as a material for boomerangs, which were both thrown during hunting and used as digging implements.

Eucalyptus oil should not be taken internally at full strength, and large doses irritate the kidneys. Always dilute it for safety. Neither should it be applied to the skin at full strength.

Euphrasia officinalis

Eyebright

Eyebright was first recorded as a medicinal herb for "all evils of the eye" in the fourteenth century, when it was mentioned as a "precious water" used to clear a man's sight. The annual, semi-parasitic herb is native to Europe and western Asia and is naturalized in the United States. It is described as semi-parasitic because it takes some of its nourishment from grass.

The common name "eyebright" derives from the appearance of the flower, which also earned the plant a place in the Doctrine of Signatures. The flower combines the colors purple, yellow, white, and red, which is suggestive of a bloodshot eye—the condition for which it has been notably used.

In Milton's epic poem "Paradise Lost" (1667), Archangel Michael offers up the herb to give Adam, father of mankind, clearer vision:

> ... to nobler sights
> Michael from Adam's eyes, the film removed,
> Then purged with euphrasine and rue
> His visual orbs, for he had much to see.

This tiny plant belongs to the foxglove family and is very attractive to bees. It is suggested that the yellow spot on the central lobe of the flower and the purple veins on either side of it help to guide bees and other beneficial insects down the plant's throat to the nectar.

- Being semi-parasitic, eyebright needs grasses as its host roots. The plant also partners with plantains and clovers, which makes it difficult to cultivate.
- It can be grown from seed scattered onto host grasses during spring. The area must remain moist for germination to take place.
- Its native habitat its sub-alpine grassland, with an alkaline soil and cool climate conditions.
- Plants are harvested while in full flower and dried.

Eyebright has ophthalmic, astringent, and anti-inflammatory properties. The herb can be infused, powdered, or made into a tincture. For centuries it has been recognized as effective for a range of eye complaints, including sties and conjunctivitis, and it also eases symptoms experienced by hay fever sufferers, especially itchy, weepy eyes. Secondarily, it is used to treat catarrhal conditions and sinus congestion, and under its Latin name *Euphrasia officinalis* it is a homeopathic remedy for colds.

There is no definitive evidence that using eyebright while pregnant or breast-feeding is safe, so it is not recommended.

According to seventeenth-century herbalist Nicholas Culpeper, eyebright can fortify the brain and memory and be used to treat vertigo.

Meadowsweet

Meadowsweet is an aromatic herb native to Europe, East Asia, and the eastern coast of North America. This hardy herbaceous perennial was an important sacred herb of the Druids and was used as a strewing herb. The scent is reputed to cheer the heart, and Queen Elizabeth I of England is recorded as having liked it placed in her chambers to freshen the air.

The herb's common name of "meadowsweet" alludes to the sweet, heady perfume of its blooms. Other common names, including "queen of the meadows," refer to the herb's tall, elegant structure, topped by delicate, lacelike flowers. In the past the flowers were added to fruit salads and fruit drinks; they were also added to mead (made with fermented honey and water) and can be used to make delicious fritters.

Placing meadowsweet on water is said to reveal information about a thief: if it sinks, the thief is a man; if it floats, a woman.

- Meadowsweet may be grown from seed, root cuttings, or division of the rootstock.
- The plant loves damp or swampy ground and is found by rivers and streams, and naturalized in ditches.
- It prefers open ground to growing in a pot.
- The plant has a long flowering period, from June to September.

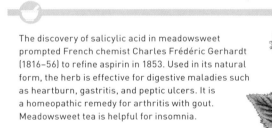

The discovery of salicylic acid in meadowsweet prompted French chemist Charles Frédéric Gerhardt (1816–56) to refine aspirin in 1853. Used in its natural form, the herb is effective for digestive maladies such as heartburn, gastritis, and peptic ulcers. It is a homeopathic remedy for arthritis with gout. Meadowsweet tea is helpful for insomnia.

Meadowsweet has a sweet taste and fragrance that lends itself to certain dishes, particularly otherwise tart fruit pies, flans, and compotes.

To aid the complexion, meadowsweet is added to base oils, waters, and creams. Use of the herb in this way is said to promote a healthy glow.

The fresh herb was once hung in bunches in bedrooms and wardrobes to impart its sweet scent. Fresh leaves and flowers may be added to bathwater for their soothing, relaxing effect.

> *During the Middle Ages, fasting pilgrims would eat fennel seeds to stave off hunger pains.*
>
> ROB LOUGHRAN, SFGATE WEBSITE

Foeniculum vulgare

Fennel

A native to Mediterranean countries, fennel was introduced to northern Europe by the Romans and to North America by early European settlers. It is now found growing wild almost worldwide, especially on alkaline soils by the sea. The herb has long been held in high esteem. In 812 CE the Emperor Charlemagne demanded "that fennel be planted on farms and in monastery gardens throughout his empire" for use as both a food and medicine.

Fennel was used by the ancient Egyptians for digestive ailments, and the ancient Greek physician Hippocrates recommended it for treating colic in infants. The Romans gave it to their warriors to keep them strong and healthy, while the ladies ate it to keep slim. Pliny suggested that it could be used to treat twenty-two different ailments.

The Roman legions of Britain left behind a legacy of "growing wild fennel" that the native peoples wholly embraced, believing the plant to have great power in healing the sick and warding off evil. In an ancient charm of nine herbs, fennel is described as "great in power" against pain and venom,"against three and against thirty," "against a friend's hand and . . . sudden trick," and "against witchcraft of vile creatures."

- Seeds are best sown in modules with a bottom heat of 59–68°F (15–20°C) for quick, even germination.
- Young plants should be planted in a sunny position, in fertile, well-drained, loamy soil.
- Fennel is a perennial plant but is likely to be killed off if the winters are wet.
- More than one plant should be grown if both leaves and seeds are required.
- The plant will self-seed and spread if left alone.

Fennel has a sweet, aniseed tang that compliments foods such as lamb, pork, oily fish, and certain vegetables, such as broad beans. Add leaves in small quantities to salads of root vegetables, potatoes, tomatoes, and beets. The herb is perfect for poaching fish and for fish stocks. Fish may be grilled, roasted, or barbecued on leaves and stalks of fennel. Add chopped fennel leaves to yogurt, sour cream, and crème fraîche for dipping and for baked potatoes. Add the seeds to cakes, buns, breads, pickles, and stewed gooseberries.

Fennel tea is good for intestinal spasms, cramps, indigestion, gas and gas pains, irritable bowel syndrome, and infant colic. Used in an eye bath, it reduces inflammation, soreness, and conjunctivitis. A fennel mouthwash is effective for gum disease.

Use a poultice of crushed leaves and stems for sore and swollen breasts due to breast-feeding. However, avoid excessive internal use during pregnancy.

Rubbing fennel leaves into the coats of pet animals such as dogs will deter fleas and lice from infesting them and their bedding. Either green or bronze-leaved varieties of fennel may be used.

Bathing in fennel has a stimulating effect on the skin. Steaming the face using a bowl of fennel mixed with boiling water will smooth out wrinkles. The seeds and leaves may also be used for facial steaming intended to achieve deep cleansing.

Wild Strawberry

The wild strawberry is a member of the rose family, and in fact its small, white flowers each resemble a tiny rose. There are a number of explanations of the strawberry's name, some of which refer to its growing habit: strawberry plants "strew" or stretch over the ground, and "strew" became "straw." Or the name may derive from the folk name of "strayberry," which again refers to the plant's natural habit.

Fragaria vesca is commonly found in woodlands and hedgerows in colder European countries such as Sweden, France, and Britain; other species of the genus occur in North America, with *F. virginiana* being indigenous to Canada and the most prolific in North America. Native Americans used different species of the plant medicinally.

The earliest record of what was called "stroeberie" is found in a Saxon plant list of the tenth century CE. By 1300, wild strawberries were under cultivation in some gardens. In 1573 English poet and farmer Thomas Tusser mentioned wild strawberry in his *Five Hundred Points of Good Husbandry*: "Wife unto thy garden and set me a plot with strawberry roots of the best to be got. Such growing abroad, among thorns in the wood, well chosen and picked prove excellent good." In the seventeenth century, street vendors, mainly women, would cry "strabery rype" as they hawked the fruit around the streets of London.

- Strawberries are easily propagated by their runners, which each have their own root system. These can be taken off plants throughout the growing season, from spring to fall, for planting into pots.
- The sprawling plants are best allotted a patch all to themselves.
- Strawberries like to grow near beans, lettuce, and spinach, but do not thrive near brassicas.
- Harvest the fruits when ready, and cut leaves early in the summer for drying.

Wild strawberries are much smaller and sweeter than cultivated ones. Both can be made into jams, conserves, cordials, and syrups. They are often used fresh to decorate cakes and tarts. Strawberries may be dried for addition to muesli and other breakfast cereals.

The leaves can be used in a decoction for their astringent properties; use as a skin toner for oily skin. Mashed fruit eases sunburn and can lighten skin. Pulped strawberries and a little baking soda may be used to remove stains from teeth (by action of malic acid) and whiten them.

People allergic to aspirin tend also to be allergic to numerous fruits and vegetables including strawberries as well as many cosmetics and medications. The compounds responsible are salicylates.

Strawberries are antibacterial, antioxidant, antiviral, astringent, digestive, diuretic, and laxative. The fruits' iron and vitamin C content makes them good for anemia and fatigue, fighting infections, and lowering cholesterol. They are also cleansing, so suitable for treating arthritis, rheumatism, joint pains, gout, and kidney stones. The folic acid they contain is helpful in preparing for pregnancy and for the first three months of it. A tea made with strawberry leaves will stimulate the liver and ease dysentery.

*One must ask children
and birds how cherries
and strawberries taste.*

JOHANN WOLFGANG VON GOETHE,
PHILOSOPHER

Galega officinalis

Goat's Rue

A hardy, perennial, herbaceous herb grown for its beauty as a plant as well as for its herbal properties, goat's rue is native to Europe. It spread from Italy into France (where it is known as French lilac), and into Germany, where the name Pestilenzkraut or "plague herb" dates from its use as a treatment for bubonic plague. In the United States it is a notifiable weed.

The English name of goat's rue may refer to the odor of the leaves and stalks of the plant when crushed, one that is reminiscent of goats. The name is also a warning to farmers that the herb is fatally poisonous to sheep and goats. Cattle and horses also find it unpalatable, but in the nineteenth century it was discovered that giving goat's rue to cows as a fodder plant could improve their milk yield by between 30 and 50 percent. The genus name *Galega* refers to this property, being derived from *gale*, or "milk," and *ega*, "to bring on."

The beautiful downward-facing pealike flowers have a mild, sweet scent, not of goat but of coconut. The flowers are either white, lilac, or pinkish-purple and appear from June to August.

The plants contain high levels of nitrogen, which makes them a valuable nutrient source for crops when ploughed back into the soil.

- Goat's rue can be grown from seed sown in early spring, or propagated by dividing the roots in fall or spring.
- Plants should be cut back in the fall or after flowering because they self-seed freely.
- As a garden escapee, the plant likes full sun and can be found growing wild everywhere, even on the central section of highways.
- In a garden it is best grown in a moisture-retentive soil.
- It attracts beneficial insects and produces a good green manure.

Dried flowering stems and leaves act as a galactagogue, that is, they promote milk flow in breast-feeding women. The plant is also diaphoretic, meaning that it induces sweating, and so can be used to reduce fever. It helps to rectify digestive problems caused by insufficient enzymes. Goat's rue should not be self-administered.

Used fresh, the herb can clot milk, and in northern European countries it is used instead of rennet for that purpose in order to make cheese.

Galium odoratum

Sweet Woodruff

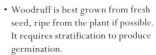

This pretty, low-growing, hardy, ground-covering herbaceous perennial is a member of the Rubiaceae or bedstraw family. It is native to northern and central Europe, the Balkans, Siberia, and North Africa, with small pockets of the herb naturalized in southern Canada and the northern United States.

The herb grows wild in shady, woody places, which explains the "wood" part of its name. The "ruff' element comes from the plant's leaf structure—neat green leaf whorls that grow around the stems like lace ruffs. The little, white, star-shaped flowers are followed by balls of tiny bristly fruits that self-seed freely in receptive environments.

In Germany, where the plant is called Waldemeister, or "master of the woods," it is traditionally steeped in Rhine wine on May Day to make the celebratory drink *Maibowle*.

Since the Middle Ages woodruff has been dried as a herb for strewing and making bedding more fragrant. As it dries, its scent of new-mown hay intensifies. The scent wards off insects and moths, and so bunches are hung in cupboards of linen and clothing. Its high level of coumarin make it useful for potpourri, snuffs, and perfumes. In perfumery, hay, honey, and vanilla may be detected in the essence and oil.

Sweet woodruff is an astringent, slightly bitter herb used for urinary and digestive complaints. It improves kidney and liver function and is used to treat jaundice and hepatitis. The leaves make a refreshing tea that has a relaxing and soothing effect on stomach pains. In homeopathy it is used to treat inflammation of the uterus. It is also a sedative and tonic.

Adding woodruff to chilled white wine enhances the flavors; in Alsace, France, it goes into the tonic white wine *Maitrank*. Flowers are added to salads and strawberries.

Woodruff products are safe to eat or drink but should not be consumed continually as they are toxic in large quantities.

- Woodruff is best grown from fresh seed, ripe from the plant if possible. It requires stratification to produce germination.
- The plant can also be grown from pieces of root. After flowering, take small pieces of root from the plant, lay them on compost in a seed tray, and cover with a thin layer of compost. Keep the compost moist, and young fresh shoots will appear. Transfer when large enough.
- The plant turns brown and dies back in fall.
- A woodland position or shady cool area is best for it to thrive.

Indian Physic

This leafy perennial herb, also commonly known as American Ipecac, is indigenous to the United States. It is found growing in humus-rich woods in the mountains and piedmonts from Massachusetts to Georgia. The plant has a light airy appearance; the stems have a reddish tinge with green leaves and star-shaped white flowers with red tingeing. Its attractiveness has made it a sought-after herbaceous border plant.

The root bark of the plant has long been associated with and used by Native American tribes. The Cherokee used it in a poultice for rheumatic pain; they chewed the root (or infused it) for bee and other stings, and it was also used as a toothache remedy. The Iroquois used it to treat diarrhea, colds, and sore throats. Some tribes used it as a powerful emetic during annual cleansing ceremonies, which is where both its common names come from. The herb was soon adopted by the early white settlers to North America and they called it bowman's root—"bowman" being their name for the indigenous people.

- Easy to grow, likes light rich moist, but well-drained, soil in partial shade and a sheltered position.
- Propagate by seed preferably sown in early fall or by division of the plant in the fall and spring.
- Plants need to be protected from slugs when young.
- Will grow to a maximum height of 3 feet (1 m).

Has various medicinal properties: emetic, respiratory aid, febrifuge, and expectorant. The reddish brown root is dried so that the bark can be removed and pounded into a powder. It was then added to alcohol to infuse the bitter constituents that contain the healing factors, turning the finished liqueur a red color. It was used in small doses to treat chronic diarrhea and constipation, and also bronchial and asthmatic conditions. Used in a poultice for leg swellings.

Caution should be taken as large doses can cause severe vomiting and purging. It has even been used to induce vomiting when required.

Ginkgo

Uniquely *Ginkgo biloba* is the only genus in its family. It is often referred to as a "living fossil" because over time its form has hardly changed as evidenced by the fossils of its leaves that have been found that predate the evolution of mammals.

It originates from China and Japan where the first seeds were gathered and sent back to Europe in the early part of the eighteenth century. They were soon cultivated so that they could be grown throughout the world. Ginkgo is much sought after for its decorative leaves, which resemble the delicate maidenhair fern.

Ginkgo trees can be found in city parks, arboreta, and gardens. Resistant to disease and insects and able to produce aerial roots and sprouts, they are long-lived, slow-growing, attractive trees that seem to cope well with pollution. The oldest recorded living tree is 3,500 years old. The trees have often been planted in temple gardens in China, Japan, and Korea.

The ginkgo tree has been used in Chinese traditional medicine for centuries and is now being cultivated for use in Western medicine, especially in treatments for dementia and Alzheimer's disease.

- Grown from ripe seeds.
- Ginkgos take about twenty years to mature, with male catkins and female flowers forming on different plants. Females bear putrid, plumlike fruits containing edible almond-sized seeds, which are highly regarded in China and Japan as a delicacy.
- If you require fruit, you need to plant one male and two females to guarantee successful viable fruits.
- Excellent urban, shade trees.

Antioxidant and anti-inflammatory. There has been much research into the properties of ginkgo, especially for memory and concentration loss, which is why it is so widely taken as a herbal supplement. The leaves have been found to hold the highest concentration of ginkgo flavonoids, which hold the key to its curative sources. It stimulates circulation and cerebral insufficiency in the elderly. It is also used for circulation problems, such as Raynaud's syndrome.

Although ginkgo has a long use in traditional medicine for blood disorders, do not use if taking an anticoagulant, antiplatelet, or any other medication for circulatory problems. Not to be used while pregnant and only under supervision of a medical herbalist.

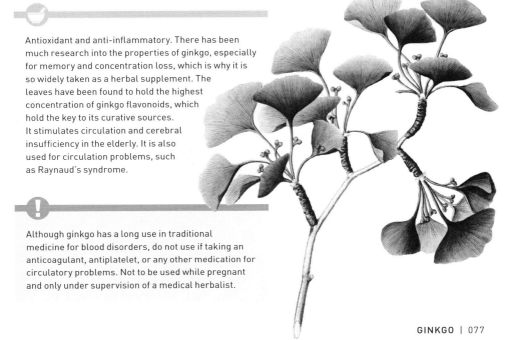

The light and dangling
licorice flowers
Gave off the sweetest smells

SIR JOHN BETJEMAN, "THE LICORICE
FIELDS AT PONTEFRACT" (1954)

Licorice

Licorice is a perennial herbaceous plant native to southeastern Europe and Southwest Asia. It is also found growing in northern China and Mongolia, and is cultivated in Russia, Iran, India, and Spain. The plant has been used medicinally for more than 3,000 years, as testified by its record on Assyrian tablets and Egyptian papyri. It was also well known by the ancient Greeks and Romans, who praised its sweetness. A piece of the root was once given to teething infants.

It was first introduced to Britain by Spanish Dominican Black Friars, who settled in Pontefract, Yorkshire, where it virtually became the main crop of the region. It was initially grown for medicinal use, when it was found to ease coughs and stomach complaints. In 1305 King Edward I ordered a tax to be placed on imports of licorice so that he could finance repairs of London Bridge.

The famous Pontefract cakes, or pomfrets, which were made from extract of licorice as early as 1614, were originally for medicinal use. A nobleman called Sir George Savile came up with the idea of applying a small stamp of the castle at Pontefract to a flat piece of hardened licorice extract. In 1760 a Pontefract apothecary called George Dunhill began adding sugar to the recipe and thus the famous Pontefract cake was born and sold as confectionery.

- Hardy perennial
- Seeds need to be presoaked in water and sown in the fall in a greenhouse.
- Cultivated by rhizome division when the plant is completely dormant. Make sure that the rhizome shows one or more eyes/buds for successful growth. A little patience is needed—they take some time to shoot.
- Plants need rich, deep well-cultivated soil in a sunny position; when first planted they will take a while to settle and grow to produce enough root to be harvested.

Used in shoe polish and soap. Most licorice is used as a flavoring agent for the tobacco in cigarettes and for chewing, although its taste is not detectable by the consumer.

Licorice is fifty times sweeter than sucrose. It is used primarily for flavoring cordials, root beers, and other soft drinks. Brewers use it to give body, color, and flavor in the making of heavy, dense, full-bodied ales, such as stouts and porters. It is also used in a wide variety of sweets and candies, although the flavor is often supplemented by aniseed oil. In Scandinavia and the Netherlands, salty licorice candy is popular.

Licorice has a long history of medicinal use: it has anti-inflammatory, expectorant, and detoxifying properties. It can be used to alleviate asthma, bronchitis, coughs, and sore throats. It is also beneficial for stomach ulcers, heartburn, and constipation. It increases bile, lowers blood cholesterol, and detoxifies the liver. Externally, it can be used to treat eczema, herpes, and shingles. It is thought to rival St. John's Wort as an antidepressant.

The compound glycyrrhizin found in licorice can cause high blood pressure and water retention. It is not suitable for people with kidney problems and should not be taken if pregnant.

Witch Hazel

A native of North America and Canada, its specific name refers to where it was first found in the damp woodlands of Virginia in the United States. Its attraction as a garden-worthy small tree are the yellow tassel-like flowers that appear in September and October in its native land, with the varieties grown in British gardens blooming in December and February. The flowers appear after the leaves have fallen, these are then followed by black nuts. When ripe and hard, these expel their seeds like the sound of a shot gun, throwing them sometimes 20 feet (6 m). This gave rise to the folk name of "snappy hazel."

The Anglo-Saxon word for bend: "wyche" or "wice" was handed down through the ages. The name "witch hazel" is believed to have derived from the tree's pliant branches being used for water divining. Colonists in North America chose the tree for seeking out places to dig for wells. If you found a "well" with your witch hazel rod, it was thought to be evidence of witchcraft, hence the expression "witching a well." Witch hazel has a long history of use as a medicinal plant by Native Americans tribes, who used it to treat various ailments ranging from colds to venereal disease. They also used it for water divining.

- Small deciduous tree or shrub with smooth brown bark; grows to 4.9 to 6.5 feet (1.5–2 m) tall. Flowers of bright yellow tassels appear after the leaves have fallen. Produces several branching stems.
- Prefers damp humus-rich soil; flowers best in an open position; can be grown in semi-shade.
- Propagate by softwood cuttings in summer. Germination can be difficult, slow, and erratic.

Witch hazel contains tannins that can help reduce swelling and fight bacteria when applied directly to skin. It has been used in cosmetics for its astringent properties to tone and cleanse.

The seeds of witch hazel are quite edible (if you can find them before they launch). They are known as hazel nuts and are said to taste something like pistachio nuts.

Witch hazel is not to be taken internally because the tannins present in the plant can cause nausea and vomiting. More seriously, it can also cause liver damage.

Witch hazel is antibacterial, anti-inflammatory, astringent, and analgesic. A bottle of witch hazel water could once be found in every pharmacy. The leaves can be used in an eyewash for conjunctivitis; to soothe wet eczema, insect bites, stings, and to cool sunburn and acne. It is ideally used for bruises and sprains, because it reduces swelling and soothes the pain. As a salve it can be used for hemorrhoids, varicose veins, scrapes, and bruises. Frozen witch hazel water is also used to reduce all kinds of swellings due to its astringent qualities.

An extract of the leaves and bark is used in the astringent witch hazel, which can be used as a tonic for oily skin, pimples, and as a cleanser. It tightens skin tissue and can be used as a tonic for puffy tired eyes. It is added to some anti-aging creams.

*Haloed with
flickerings
of yellow blaze
. . . wise in spells of
drugs and gums*

ELIZABETH AKERS, "WITCH
HAZEL," *CENTURY MAGAZINE* (1894)

> *Napoleon . . . famously once remarked that he could smell the island of his birth [due to the curry plants on Corsica] from his exile on Elba.*
>
> ODA O'CARROLL ET AL., *CORSICA LONELY PLANET GUIDE* (2004)

Curry Plant

As its Latin species name suggests, *Helichrysum italicum*—commonly known as the curry plant because of the strong smell of its leaves—is a native of Italy and other southern European countries. It grows on dry, rocky, or sandy ground in the Mediterranean region. It is a flowering plant in the daisy family, Asteraceae. Very little is known about the history of the plant's origins and uses, and it is considered a relative newcomer to herbal lists in comparison with other aromatics. Its attraction lies in our insatiable love of all things spicy, especially foods such as curry. Although the leaves smell currylike, the plant has nothing to do with dried curry powder or the curry tree.

The curry plant has dense silvery foliage of needle-shaped, hairy leaves and small yellow flower heads that appear in summer. It has an upright shrubby habit, which becomes more sprawling with age, reaching a height of 2 feet (60 cm). It is a good companion plant for sages, especially the purple and gray-leaved varieties, which form attractive partnerships and displays.

H. italicum, the dwarf *H. italicum* subsp. *microphyllum*, and the cultivar 'Darlington' all make good hedging and edging plants because they can be controlled and clipped, and they are an excellent choice for edible gardens.

- Needs a well-drained, sunny position in a neutral to alkaline soil to thrive.
- Hardy perennial that only suffers when conditions are exceptionally wet in winter, when it can rot.
- Cut back after flowering; leave at least 2 to 3 inches (5-7 cm) of soft growth to maintain health and good shape; do not cut right back because it will kill the plant.
- Propagate by soft cuttings in the spring and semi-ripe in the fall.
- Protect if temperature drops below 14°F (-10°C).

The curry plant is used in Mediterranean meat, fish, and vegetable dishes. The young shoots and sprigs of leaves are infused in the dish while cooking to impart flavor and then removed before serving. Add a few leaves or sprigs to vegetables, rice dishes, mayonnaise, creamed eggs, and cream cheese. The leaves smell stronger than they taste—they have a resinous, bitter aroma similar to sage or wormwood —making their use in cooking limited.

The herb has anti-inflammatory, fungicidal, astringent, and antiseptic qualities. The oil from the blossoms is used medicinally to soothe burns, chapped hands, and sports injuries.

The yellow flowers have an intense fragrance and can be used in potpourri and in dried flower arrangements. Oil produced from the flowers is also used as a fixative in perfumery.

Some gardeners have reported that planting bushes of curry plant in the garden can be effective as a natural cat deterrent. Cats dislike the plant because of the odor it gives off; the plant's coarse texture is also said to irritate them if they brush past it. Try scattering bruised clippings of the plant around their visiting areas. Alternatively, make an infusion of the herb combined with others with similar properties, such as lemon balm, to use as a spray.

The culture of the hop . . . so analogous to the culture and uses of the grape, may afford a theme for future poets.

HENRY DAVID THOREAU, *CIVIL DISOBEDIENCE AND OTHER ESSAYS* (1849)

Hop

Hops are a climbing, hardy, herbaceous perennial that from ground to maturity can reach 10 to 20 feet (3–6 m) in one season. The plant is native to Europe, Asia, and North America and is widely cultivated in the warmer parts of the northern hemisphere and South America. It was first noted by Pliny in the first century CE, when he recorded hops being grown as a popular garden plant and vegetable. He mentions hop shoots being gathered in the spring to be sold in markets and then eaten like asparagus. Hops were first used for brewing in ancient Egypt.

The Anglo-Saxon word "hoppan," meaning "to climb," is thought to be where the common name originates, whereas the generic name comes from "humus" ("earth") because the plant needs rich soil to thrive. The interesting specific name "lupulus" ("wolf") refers to the growing habit of the plant, which "strangles" as it climbs like a wolf strangles its prey.

The practice of using hops to flavor ale was introduced to Britain by immigrants from the Netherlands, where their ability to flavor and preserve saw them take over from alecost and other plants. King Henry VIII forbade their use and instructed parliament to petition against "the wicked weed" because it would spoil good ale. Nonetheless, hops soon became the chosen flavoring for brewing and they are still cultivated in many counties in England today.

- Needs rich deep soil in a sunny position; tolerates some shade.
- Needs to be kept watered in dry spells with a good mulch.
- Can grow 6 inches (15 cm) in a day; needs trellis or an arch to grow up and over. Twines in a clockwise motion.
- In hop yards only four of the shoots are needed to climb the strings, which are manually encouraged by "tying." The hops then climb on their own to the tops of the wires where they will produce their hops. Only female plants produce hops.

Young shoots are slender and look like asparagus spears; they can be steamed and eaten with butter as asparagus. They can also be used in savory flans, quiches, soups, and salads.

Hop pillows stuffed with freshly dried hops or made into sachets soothe and aid sleep. Warmed hop pillows can be used to treat earache and alleviate tension headaches.

Brown dye is produced from the leaves and flowers. The stripped stems are strong and pliable enough to weave into baskets and mats. Pulped hops can be made into paper.

Not to be taken in large doses or combined with other sedating herbs or medication because of drowsiness. Do not take if suffering from depression or taking anti-anxiety drugs.

Hops have some medicinal use; the strobiles (female flowers and cones) are collected and dried. Use only freshly dried green hops. They are antimicrobial, digestive, diuretic, and sedative. Their sedative properties are due to the presence of compounds that depress the central nervous system. An infusion of the flowers can aid digestion, improve appetite, soothe colic, and help ease irritable bowel syndrome. Note that infusions will be bitter if left to stew.

Hypericum perforatum

St. John's Wort

Hypericum perforatum, commonly known as St. John's Wort, is a perennial native to Britain, Europe, and Asia that has naturalized in North America and Australia. A quickly spreading plant, the genus name comes from the Greek meaning "over a phantom" or "apparition," which refers to its repugnant scent driving away evil spirits. The species name refers to the strong ribbed leaves that when held up to the light reveal clearly defined dots as if perforated. Its common name refers to St. John the Baptist because the herb came into full flower on June 24, the feast day of St. John. This was thought to be an auspicious day to harvest the herb for medicinal use; it was thought safer to collect the herb early in the morning while it was still wet with dew and before the sun had risen. "Wort" is an old Anglo-Saxon word for medicinal herb.

The whole plant has been known and used by many cultures as a plant of protection from malevolent forces. Hanging bunches of the plant above the door of your house was thought to give protection from thunder, lightning, and fire; it also kept at bay any passing witches and stopped the devil crossing the threshold. Scots wore a sprig to protect against the "evil eye."

In England as well as in New England in the United States, St. John's wort was once used to wrap cheeses for transporting by land and sea because it helped preserve them.

- Seeds are tiny and can be sown in the spring into modules or seed trays, covering lightly with soil.
- Self-seeds prolifically once established. Cut back after flowering to prevent this happening.
- Happy in most soils, but especially rampant on sand.
- Grows wild in many meadows.
- Although grown commercially in parts of Europe, St. John's wort is listed in more than twenty countries as a noxious weed.
- Yellow flowers appear from late spring to midsummer.

Dye is extracted from the root and buds. Brown or red is the most common color, but there are varying degrees, including yellow and black, dependent on mordants (to fix light fastness) used.

The medicinal properties of St. John's wort are antibacterial, anti-inflammatory, antiseptic, antiviral, nerve tonic, pain relief, and sedative. Macerating the flowers and leaves in a good quality base oil produces a reddish colored oil that can be used for painful joints, neuralgic pain, rheumatism, arthritis, bruises, and strained muscles. However, it is most widely used in commercial preparations to treat depression and is known as natural Prozac.

It should be eradicated from livestock grazing fields as it can become a cumulative poison. The red dye pigment in it can also photosensitize grazing animals.

A vigorous spreading plant that produces up to 30,000 seeds per plant each season. Plant it where it can be watched over, so as not to become a nuisance.

Generally well tolerated, but overuse can make the skin sensitive to light. Seek medical advice before taking and do not take with any other medication.

St. John's wort doth
charm all witches away,
If gathered at midnight
on the saint's holy day.

OLD ENGLISH POEM (1400)

> *He spoke of trees, from the cedar that is in Lebanon to the hyssop that grows out of the wall.*
>
> THE BIBLE, 1 KINGS 4:33

Hyssop

Hyssop is a bushy, hardy, semi-evergreen perennial native to Mediterranean regions and the Middle East, where it was highly esteemed for its cleansing, medicinal, and flavoring properties. It can be found growing wild on dry scrub, in old walls, and around derelict buildings. This aromatic herb has been cultivated in gardens for more than 500 years. It was taken to the New World by settlers to use as medicine, be drunk as tea, and smoked as a herbal tobacco.

The Bible refers to it in relation to purification; in the Book of Psalms, David asks God to spiritually cleanse him: "Purge me with hyssop and I shall be clean; wash me, and I shall be whiter than snow." It was also used in the cleansing of lepers and leprous houses, although experts do not agree as to whether this was the biblical hyssop. In fact, arguments have ensued for centuries between botanists over this particular plant. It is known that it was used in the consecration of Westminster Abbey. It was also a favorite edging plant in Tudor knot gardens and was used as an Elizabethan strewing herb.

Hyssop is undervalued today in comparison to its historical use, but its attractive habit and pretty flowers make it excellent for edging pathways. Rock hyssop (*Hyssopus officinalis* subsp. *aristatus*) has a more compact dwarf habit, and is particularly suited to this purpose.

- Can be grown from seed. Seeds are small so sow into modules or trays in early spring under glass, with bottom heat (60-70°F/ 15-21°C) for germination.
- Take cuttings late spring/early summer from new growth on non-flowering stems.
- If growing a hedge, trim after flowering to maintain the shape and vigor.
- Prefers a well-drained, sunny spot.
- Deep blue-purple, tubular flowers from June to August.
- Harvest the whole plant and dry for winter use.

Excellent for attracting wildlife, particularly bees. It deters cabbage white butterflies from brassicas. Beekeepers love hyssop because it produces a rich aromatic honey.

Due to its strong flavor, use sparingly in cooking. The leaves have a slightly bitter taste and a minty aroma. They make a refreshing tea. They are often added to fatty foods as they help with digestion. Chop a few leaves into stuffings and sausages. Fresh flowers and the tender tops of the plant can be sprinkled into salads. A good partner with sweet cicely (see p. 121) in stewed fruit and fruit pies. It is also used for flavoring some liqueurs, especially Chartreuse.

As a medicinal herb, hyssop has antiseptic, antiviral, sedative, and expectorant properties. It can be added to bathwater to soothe, relax, and ease aching limbs. Make it into oils, salves, and tinctures to treat rheumatic pains. Gargle as a mouthwash for sore throats and to ease gum and tooth infections. The leaves can also be infused and taken to treat digestive problems, break a fever, and expel catarrh.

The oil is banned in some countries as it can cause seizures. Avoid if you have thyroid problems or a heart condition due to its high iodine content. Do not take if pregnant or breast-feeding.

The holly and the ivy,
When they are
both full grown,
Of all the trees that
are in the wood,
The holly bears the crown

TRADITIONAL CHRISTMAS CAROL,
"THE HOLLY AND THE IVY"

Holly

Holly is native to Europe and can be found growing as far north as coastal Norway. It is often found growing in the shady undergrowth of forests, particularly oak and beech, and in hedges. It is also commonly grown in parks and gardens in temperate regions.

Holly is a legendary herb whose use dates back to pagan times, when it was used as protection against hazards as varied as evil spirits, poison, lightning, and wild animals. It was known as a good-luck herb for men, as ivy was for women. Druids were the first to decorate their homes with holly. Like mistletoe, it carried the mantle of all that was good due to the bright green leaves and red berries, which symbolize life in the middle of winter when almost everything else is dormant. The red berries were thought to give protection from witches and witchcraft.

The Romans used holly for celebrations, such as the festival of Saturnalia in December, when gifts decorated with holly boughs were sent to friends. Some of these customs were later adopted by Christians during their Christmas celebrations. Traditionally, Christmas pudding is still decorated with a sprig of berried holly as it is brought to the table. The most-documented legend is that holly first appeared under the footsteps of Christ; its thorny leaves and red berries symbolized his suffering and wounds, which is where the common name of "Christ's thorn" originates.

- Evergreen bush or tree that grows to 10 to 50 feet (3-15 m); slow-growing with no preference to soil type; it has a great capacity to adapt to different conditions, but given a rich gravelly loam it will reach its full potential.
- Sow seeds; from green berries will germinate in one year; from red it takes two years.
- Transplant young plants in the fall when they reach 18 inches (45 cm); place in pots or humus-rich soil.
- Dark green, leathery, shiny leaves with sharp spines.
- Makes a good hedging plant.

Holly wood is hard, close-grained, and almost white in color, which makes it ideal for white chess pieces (ebony wood is used for the black). It is also used for inlaid marquetry and turned to make walking sticks and tool handles. It was used in the 1800s to make the spinning rods for looms, but was eventually abandoned due to the density of the wood wearing and snagging the threads. Some musical instruments are made from holly wood.

Holly is a reliable stock-proof hedge plant, which provides shelter in the winter and shade in the summer. The foliage has also been used for animal fodder. The prickly leaves provide nutritious food for deer in the depths of winter. The flowers produce much-needed early spring nectar for honeybees and butterflies. The plant is also important to bird life as the berries are plentiful at a time when most other sources of food are in winter hibernation.

Holly is seldom used medicinally, but it is diuretic and diaphoretic. Only the leaves were used, collected in May and June, and infused for coughs, colds, flu, catarrh, bronchitis, and pleurisy. Used in Bach Flower Remedy for oversensitivity.

Berries contain alkaloids, caffeine, and theobromine and are regarded as poisonous, so do not eat.

Indigo

Indigo is native to India and is cultivated mainly in subtropical countries. It is classed as a sub-shrub, which belongs to the pea family, the Leguminosae. Evidence of the blue dye from indigo has been found on pieces of cloth in ancient Egyptian tombs, indicating its use for at least 3,000 years. When trade routes from Europe to the East Indies became established in the sixteenth century, indigo was one of the many commodities brought back to Europe, where it was enthusiastically embraced and came to replace native woad (see p. 95) as the source of blue dye. It was in great demand in the European textile industry. By the eighteenth century, it had found its way to the North American colonies, where wild indigo (*Baptisia tinctoria*) and blue false indigo (*B. australis*) had been the source of a much weaker blue dye. In South Carolina indigo was cultivated on a large scale. In 1859 farmers in Bengal, India, revolted against growing indigo. True indigo dye was not readily available; rich, deep, and with a better natural color fastness, it takes several processes of fermentation to produce the depth of color. Huguenot planters continued to grow indigo for domestic use after it collapsed as a crop in the United States.

- A perennial in hotter climates/annual where it is cooler. Needs protection and warmth if grown in temperate climates.
- Can be grown from seed; soak the seed for twelve hours in hot water to quicken germination. Sow into modules under glass with bottom heat in the spring.
- Can also be grown from cuttings when the plants start to "grow away" in late spring.
- Grows well in nutrient deficient soil.

The best plant for the best brilliant blue, indigo is still used by specialist cloth and clothing producers, natural dyers, and designers. It is now commanding high prices for the garments produced. It was used to dye denim jeans before synthetic substitutes.

People have used indigo, alone or in combination with henna, to dye their hair black for about 4,000 years.

Indigo is no longer used in traditional medicine because it can cause severe vomiting.

Not a chest of indigo reached England without being stained by human blood.

COMMISSION REPORT ON
INDIGO REVOLT OF 1859

Elecampane

Elecampane is considered one of the largest perennial herbaceous herbs. A member of the daisy family (Asteraceae), it originates from Asia and is found in China, Mongolia, and Korea. It traveled to Europe, where it was specially prized in Holland, Switzerland, Germany, and France for its essential oil. At one time it was used to flavor absinthe. It was introduced to North America by settlers and is now naturalized from Nova Scotia to North Carolina and westward as far as Missouri.

The root was eaten by ancient Romans as a vegetable and according to Pliny, the Empress Julia Augusta "let no day pass without eating some of the roots candied, to help the digestion and cause mirth." The name *helenium* gives another possible explanation for its etymology. Helen of Troy was said to have been holding a bunch of elecampane when she was taken away by Paris; another legend holds that the plant grew from the tears she shed. Apothecaries in the Middle Ages candied the roots into flat, pink sugary cakes, which they used to alleviate asthma and indigestion. Elizabethans made a confection similar to marzipan from its ground roots mixed with eggs, sugar, saffron, and other spices. Scandinavians put an elecampane flower in the center of a nosegay to symbolize the sun and the head of the Norse god Odin.

- Propagate by seeds, which appear on the plant after flowering; they look similar to those of the dandelion and float on the air in the same way. Sow into seed trays or plant into pots or the garden when large enough to handle.
- Root division of mature plants in the fall or by off-shoots around the base. Self-seeds freely: cut off seed heads when they first appear.
- Dig roots in the fall of the two-three-year-old plants; can be frozen immediately or dried. Collect flowers when they first open for using fresh or dried.

Not to be taken by those allergic to the daisy or sunflower family of plants. Can cause intestinal irritation. Do not take in large quantities and consult a qualified herbalist practitioner before use. Avoid if pregnant.

The root was used in ancient medicine for its antibacterial, antifungal, anti-inflammatory, antiparasitic, antiseptic, diuretic, expectorant, and immunostimulant properties. A decoction of the root is best mixed with honey for chest and respiratory complaints; it soothes and reduces coughing, and loosens phlegm. Good for chest infections, asthma, and breathing allergies, such as hay fever. It can also be made into tinctures and syrups for storing.

Iris germanica var. *florentina*

Orris

This beautiful plant is a native of the eastern Mediterranean and is grown ornamentally and specifically for its violet-scented rhizome. In spring this bearded iris produces white flaglike flowers with yellow throats and white falls. The Greek word *iris* (meaning "rainbow") describes the multitude of colors found in this plant family. There are hundreds of hybrids with colors ranging from white to jet black. The common reference is directly from the word "iris" becoming "orris." The dried powdered rootstock has been used in perfumery since ancient Greek and Egyptian times. It was also once used to sweeten the breath in the form of tooth powders and lozenges.

The Florentine part of the name is associated with the early Middle Ages. Its cultivation and use was recorded by the Bolognese writer on agriculture, Petrus de Crescentiis, in his thirteenth-century *Ruralia commoda*. The plant remains the emblem of Florence (fleur-de-lis), appearing on its heraldic arms, and is still found growing in the region.

- Hardy perennial.
- It can be grown from seed, but it is much quicker and easier to divide the rhizome roots in late spring or early fall.
- Prefers a well-drained rich soil in a sunny situation. When planting irises, always make sure part of the rhizomes is exposed.
- If you want to harvest the root, allow the plant to mature until at least two years old. Dig up the rhizomes in the fall for drying and storing over winter.

Once used to treat digestive disorders, flatulence, heartburn, nausea, and headaches associated with the stomach. However, it is now mostly valued for its perfume.

It is used as a fixative in potpourri to prolong the overall perfume. It is also put in sachets to scent linen cupboards.

It can be used as a base for dry shampoos and as an ingredient in natural face masks.

Beware if handling large quantities of orris roots. They can cause nausea, vomiting, and facial neuralgia.

Woad

Woad is a biennial or short-lived perennial dye plant native to central and southern Europe and Asia. It has been in cultivation since ancient times, and has become widely distributed around other parts of the world, notably North America. There are, however, restrictions on its growth in parts of the latter due to its invasive habit.

Woad has been a staple source of blue dye in cold climates for thousands of years. Ancient Britons are believed to have painted their bodies with woad before going into battle. This intimidated their opponents, while also protecting their bodies from infection. Woad needs to be fermented to release its blue dye, which in turn gives off a foul smell. Queen Elizabeth I found it so disagreeable that she banned its production within five miles of any of her palaces, even though it was a highly important commodity at the time. In France woad was in continuous cultivation from the thirteenth to eighteenth centuries; in the sixteenth century there were 220 master dyers in Paris alone. Woad was demoted to second place when indigo (see p. 92) arrived. The last two woad mills in Lincolnshire, England, stopped turning in the 1930s. Today woad is only used by craft dyers.

The herb is so astringent that it is not to be given internally as a medicine. Only use externally and under the supervision of a qualified medical herbalist.

Historically, woad was used as a wound herb to prevent infection due to its astringent properties. It was also used as a plaster to staunch bleeding in battle.

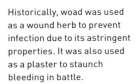

The leaves of twenty-four plants can dye 113 grams (4 oz.) of wool. Pick leaves from rosettes of plants during the first year, any time from midsummer to autumn. Once frosted, woad plants lose their depth of colour and are of no use in the second year. Consult a natural dyers' reference book for the process of dyeing with woad. The quality of the resulting dye depends on the correct process.

- Sow from seed each year to ensure a constant crop of the leaves to yield the blue dye.
- Bright yellow flowers appear from June to September followed by what look like black dangling keys, which are the seeds.
- Grows most happily in good well-drained soil with a sunny aspect. Self-seeds freely. Cut off some of the seeded heads if you want to control its spread.
- Pick leaves before the plant comes into flower.

But he himself went a day's journey
into the wilderness, and came and
sat down under a juniper tree

THE BIBLE, KINGS, 19:4.

Juniper

Juniper is a spreading shrub or small evergreen tree with green needlelike leaves, which produces berrylike seed cones. It is native to northern Europe, Southwest Asia, and North America, where it was highly regarded by Native American tribes. The Blackfoot used the berries to treat pulmonary problems and venereal disease. The Cheyenne burned the leaves as incense in a ceremony to allay fears of thunder, whereas the Potawatomi made a compound containing the berries to alleviate urinary problems. The Navajo used it as a talisman for good luck and as a smoke for hunters.

In ancient Greek and Roman times, juniper was used both for its medicinal potency and as a flavoring in food. It was strewn as an aromatic to sweeten the air in rooms; it was also burned in sick rooms and in the streets in times of epidemics to supposedly purify the air and halt infection. French hospitals still used this method of fumigation during the smallpox epidemic of 1870.

Juniper is probably best known as the predominant flavoring in the alcoholic drink gin. The word "gin" is derived from the Dutch word "genever." Oil distilled from juniper berries flavored Holland or Dutch gin, from which gin as we know it today evolved.

- It is a hardy perennial grown from seed or semi-hardwood cuttings.
- Likes dry soil in an open exposed sunny position.
- Berries will only be produced if both female and male plants are grown.
- Berries only grow on female plants, they ripen individually with various stages of ripeness occurring on the same plant. It can take up to three years to ripen turning from green to blue-black. The flavor is more intense in warmer climes.

The berries are usually crushed before use to release their bitter, spicy flavor. Dried berries and oil are used in the distilling industry to flavor gin and other spirits. Use the dried berries in marinades, especially for roast meats and game, as well as in savory jellies and conserves to serve with cold meats. Juniper wood can be used to smoke and cure meat.

Juniper has a long history of medicinal use. It is anti-inflammatory, antioxidant, antiseptic, antiviral, and diuretic. Use berries in tea to treat rheumatism, arthritis, urinary complaints, bronchial catarrh, and fevers. Also good for loss of appetite and digestive ailments. Apply as a compress to soothe aching joints. Historically, berries were thought to increase male potency.

According to folklore, juniper was hung at the door to protect against theft and evil forces. Burned juniper was used to purify temples in ancient times. It was also used to break curses.

Even though ripe juniper berries are blue-black in color, using them to make a dye will naturally yield shades of brown or khaki. Use mordants of alum or chrome.

Not to be taken during pregnancy or if you have kidney problems. The use of juniper oil, when distilled from the berry, not the wood, (it can be made from both) is prohibited in some countries.

Neither witch nor devil,
thunder nor lightning,
will hurt a man where
a bay tree is.

NICHOLAS CULPEPER, *COMPLETE
HERBAL* (1652)

Bay

Bay is an aromatic evergreen shrub or small tree that is native to the Mediterranean, but can be found naturalized all over the world. Although it is occasionally found growing wild, it is mainly cultivated in gardens. Its specific name of *nobilis* points to its status in ancient times as a symbol of peace, honor, and victory. Wreaths of bay crowned the heads of emperors, soldiers, and poets—hence the title "poet laureate" being awarded for academic excellence.

At Delphi in ancient Greece, the oracle spoke of her prophecies with a leaf of bay held between her lips. It is possible that the narcotic properties of bay were used to induce trances. In Greek mythology, when Apollo, the god of light, music, poetry, and prophecy, tried to seduce the nymph Daphne, her father, the river god Peneus, turned her into a bay tree to protect her. Racked with guilt at his actions, Apollo from that day forward always wore a wreath of bay.

Bay is considered one of the most important purification herbs and has a long history of use as a "burning" herb by both pagans and Christians to repel evil and sickness. Its branches were used to sprinkle water during ceremonies for cleansing and purifying and it was also used as a strewing herb in Elizabethan England. Today, however, the bay leaf is most commonly used as a seasoning in cooking and as an aromatic ingredient in cosmetics and colognes.

- Perennial with aromatic glossy green leaves; small pale yellow waxy flowers in spring. Green berries turn black in the fall.
- Difficult to grow from seed in cooler climes.
- Cuttings in late summer are not easy to strike; takes time without a propagator system. Take several slips to counteract losses.
- Take offshoots from the tree base; once rooted, transfer. Keep in a pot for at least a year so that roots develop.
- Can be grown in a pot or open ground. Protect for first three years in cold climes.

Traditionally, bay trees were once cultivated outside the door of the house to bring good fortune and ward off harmful spirits. Bay was added to potions designed to enhance psychic powers and it was thought that placing a bay leaf under the pillow would induce prophetic dreams.

Extracts of bay are used in traditional medicine: analgesic, antispasmodic, and antiseptic. Oil of bay is used for relief from arthritic pains, lower back pain, muscle spasms, and sprains. It also aids digestion. The essential oil is also used in massage therapy and in aromatherapy.

Widely used in cooking, especially in meat casseroles, preserved potted meats, patés, stocks, and marinades. It is one of the main ingredients of bouquet garni. The whole leaf is usually used during cooking and removed before serving. Bay is also good in fish dishes, such as shellfish soup. Bay wood makes a sweet smoke for smoking meats; burn the leaves when barbecuing meat. Bay can also be used to flavor homemade vanilla custards and vinegars.

Some people are allergic to bay leaves. Contact dermatitis and asthma have been reported. Use with caution and in small amounts. Consult a qualified medical herbalist before use.

Lavender

Lavender's origin is mainly in the mountainous regions of the southern and western parts of the Mediterranean, where it is still found growing wild. However, its universal popularity and its multitude of uses has led to it being cultivated all over the world.

Lavender is thought to have been named from the Latin word *lavare*, meaning "to wash." The Romans, who relished their baths, used the plant to perfume their bathwater. They extolled the virtues and therapeutic properties of the herb as an antiseptic and disinfectant. Indeed, it was the Romans who are thought to have introduced lavender to Britain, where it continued to be cultivated in monastic gardens for its valuable healing properties. Lavender is synonymous with an English garden and some of the finest oil comes from lavender grown in Norfolk. However, it is in France, where lavender has been grown on a large commercial scale since the seventeenth century, especially for perfumery. Provence is world famous for its breathtakingly beautiful fields of lavender.

Ancient Druids were fond of adding lavender flowers to their love potions and they burned branches of it during childbirth to cleanse the air, calm the mother, and bless the baby. In medieval times it was used for treating head lice and was also a herb used during times of plague to mask the smell of death and also to protect against infection.

- Hardy evergreen perennial. The whole plant is highly aromatic.
- Mauve-purple flowers on long spikes appear from midsummer.
- Propagate by cutting in spring or fall from the new strong growth. Once rooted, transfer them; leave until they are well grown in the pot before planting them outside.
- Harvest flowers just before they open. Hang small bunches in a warm place to dry; keep out of direct sunlight to maintain a strong color and scent.

The flowers and leaves are a good flavoring in shortbread and cookies, but use sparingly or it will be overpowering. Also good in jellies, jams, vinegars, and oils. An ingredient in the French *Herbes de Provence*.

Flowers and leaves are used in herbal medicine: antibacterial, antifungal, antimicrobial, antiseptic, sedative, and for pain relief. Infuse the flowers (often with other herbs) to treat nervous headaches, induce sleep, reduce anxiety, calm irritability. Oil of lavender is used for massage to soothe the strained and tense muscles, aches, and pains; it is also good for rheumatism and lumbago. Add to a salve for cuts, blisters, rashes, athlete's foot, burns, and insect stings. Lavender is also one of the top five aromatherapy herbs.

Lavender is one of the most popular cosmetic herbs that has long been used to perfume many products from skin tonics, cleansers, and lotions to hand and foot creams. It has also been used as a hair tonic to treat dandruff.

Lavender is one of the most popular garden plants for attracting bumblebees and honeybees, which is important when many bee species are under threat. Butterflies love it, too. It is also a mosquito and ant deterrent.

Lavender is for
lovers true,
Which evermore
be faine,
Desiring always
for to have,
Some pleasure for
their paine.

ATTRIB. CLEMENT ROBINSON,
A HANDEFULL OF PLEASANT
DELITES (1584)

Let us lodge among the henna shrubs . . . Let us see if the vine has flowered, If its blossoms have opened.

SONG OF SOLOMON, SONG OF SONGS, OLD TESTAMENT, 7:11–3

Henna

Henna is native to Egypt, Persia, Syria, India, and Kurdistan. It is also naturalized in tropical America and has been introduced to Australia. In India, Egypt, China, Morocco, and Iran, it is cultivated commercially for its leaves. Henna, *al kenna* in Arabic, has been held in high regard throughout history, especially in Arabic and Indian cultures. The red coloring produced from the leaf was thought to represent the fire and blood of the earth, linking humans to the natural world. Henna has been used as a cosmetic hair dye for more than 6,000 years and it has also long been the natural dye of choice for adorning and decorating nails, hands, and feet. It is used in *mehndi*, the henna paste used for temporary tattoo decoration by the Hindu community. The Berbers of North Africa also use it to color corpses and young babies and to create elaborate designs for marriage ceremonies.

In Europe, the women connected to the nineteenth-century Pre-Raphaelite group of artists in England used henna, although Gabriel Rossetti's wife and muse, Elizabeth Siddal, who features in so many of his paintings, had natural flowing red hair.

Henna continues to be used as a hair dye to the present day, although it passes in and out of fashion. The best powdered henna for giving rich deep red tones to hair is sourced from Persia.

- Shrubby perennial plant with greenish brown leaves; small highly scented white, light red, or deep red flowers, are followed by round blue-black berries.
- Can be grown as an ornamental, but needs constant warmth and good light levels to thrive.
- Likes well-drained sandy soil in full sun. Minimum temperature 50°F (10°C).
- Sow seed in spring and take softwood cuttings in winter.
- Harvest young leaf shoots throughout the growing season; dry and powder.

It can cause allergic reactions, especially with sensitive skin. Some pre-mixed henna body art pastes have ingredients added to them to darken the stain, which can also irritate skin.

Historically, henna has been used as an astringent and stimulant. Recent research has found that henna contains antibacterial, antifungal, and antiparasitic properties. It was used as a folk medicine in Africa to treat leprosy and amoebic dysentery. It was also used to treat jaundice, smallpox, skin infections, wounds, ulcers, and gonorrhea. A paste of powdered henna leaves was applied to the forehead to relieve headaches and fever.

Henna is mainly used for adornment. Most of the squares in Marrakech, Morocco, have henna artists ready to apply their designs to hands and feet. For this purpose, the leaves first have to be dried, milled, and sifted to make a fine powder. This is then mixed with either lemon juice, strong tea, or another acidic liquid to make a paste the consistency of toothpaste, which is used to make the finely detailed body art. It produces a rich red brown color.

Use of henna is restricted in some parts of the world. The Food and Drug Administration (FDA) in the United States has not approved henna for direct application to the skin.

Lovage

Lovage is a tall perennial plant native to the Mediterranean region. It is found in abundance in Liguria, Italy, where its name is likely to have originated from the Latin word *ligusticum*. Lovage has spread throughout temperate regions of the world, including North America, where it has naturalized. In Europe the leaves are used as a herb and the root is eaten as a vegetable.

It was grown by the ancient Greeks and Romans for both medicinal and culinary purposes. It was also a favorite plant in monastic gardens where it was grown for medicinal use. A popular restorative cordial was once made from lovage, with additions of yarrow (see p.10) and tansy (see p.196), that was given as a tonic and diuretic. Its most popular use was as a love potion herb used for aphrodisiacs and love charms, hence its common name "love ache" (a medieval name for parsley). Its properties as a deodorizing herb may have made a person more appealing in an age when people rarely washed. Lovage leaves were also placed in the boots of travelers to absorb the moisture and odors. In New England the root was candied and used as a lozenge for bad breath.

- Hardy perennial that grows to a height of 6 feet (2 m) and a spread of 3 feet (1 m).
- It can be grown from seed; sow in spring in plug trays. Transfer when roots are developed well. Grow on until the plant fills the pot, then plant outside.
- Prefers rich soil, moist but not waterlogged because this will rot the roots. Full sun or partial shade.
- Takes three to five years to reach maturity.
- Harvest leaves before flowering—they get bitter.

Lovage may increase the risk of bleeding. It is not to be taken by those who are pregnant or have kidney or bleeding disorders.

Lovage is a digestive and diuretic. Use leaves in a bath to soothe skin irritations. A decoction of seeds or infusion of leaves can be used for digestive ailments, kidney stones, urinary tract infections, and painful menstruation. Also good for increasing perspiration and easing coughs. Use in an eye bath for sore eyes. Gargle for throat and mouth infections. Syrup for chest complaints and coughs.

Lovage's natural salty stock taste makes it good in soups, casseroles, and sauces. Excellent in an herb butter. Use young fresh leaves in salads. Add to wilted greens, such as spinach, chard, or arugula. It is also good in egg and potato dishes.

Ligusticum scoticum

Scottish Lovage

Scottish lovage is a hardy perennial that grows on the cliffs and rocky shores of Scotland, Northumberland, and throughout northwest Europe from Denmark to Norway. It seems to have originally been a cultivated garden plant that escaped into the wild where it now grows abundantly.

At one time it was a monastic potted herb, which was cultivated in a similar manner to celery. All parts of the plant are aromatic when bruised. The folk names of sea lovage and sea parsley stem partly from where it was found growing and its association with sailors. Because it has a high vitamin C content, sailors would often eat the plant when they returned from long voyages suffering from scurvy.

In the Scottish Hebrides, the leaves have been used as a vegetable, boiled as greens, or commonly used fresh in salads. Known there as "shunis," the herb traveled across the sea to the Atlantic coastline of North America, where it has associations with Native Americans who liked to eat the peeled stems raw.

- A much stouter and shorter plant than lovage, with branching deep red stems, shiny three-lobed leaves, and umbels of white, pink-tinged flowers, which bloom in June and July.
- Perennial plant grows to a height of 2 feet (60 cm).
- Grow from fresh set seed for the best results, straight from the plant.
- In a garden situation, it prefers damp semi-shade.

Diuretic, calmative, and digestive, Scottish lovage was once used to treat rheumatism and digestive problems, especially gas. It was recommended as a stimulating tonic and an aphrodisiac. The seeds were ground and used to improve the taste of bitter medicines.

This plant loves the coast so much it is often found in the rocks splashed by the sea.

WWW.UKWILDFLOWERS.
COM

The stems and leaves of Scottish lovage were once given to cattle as a treatment for worms.

Use young leaves, stems and fresh green seeds in soups, casseroles, with fish and certain game.

Linum usitatissimum

Flax

Flax (also known as common flax and linseed) is thought to have originated in India and was used throughout the ancient world for its natural fibers to produce cloth. Archaeologists have found evidence of flax fibers in Syria, Iraq, and Iran, dating back to 8,000–6,000 BCE. Ancient Egyptian tombs contained illustrations on the walls that depict the spinning of flax fibers and mummies have been found wrapped in cloth made from wild flax. The Latin name means "most useful plant" as it clothed, fed, and healed those that grew and harvested it. Flax fibers are used to make linen, which was cultivated in Ireland from 500 CE.

Flax is cultivated throughout the world for various uses: fiber from the stems; seeds for human and animal consumption as a highly nutritious omega rich food; and for its valuable oil, which is gleaned from the seeds and is used for food, animal fodder, and in paint and wood finishing products.

Artists' canvases were once made of linen. It was also used for sailcloth, awnings, and horse rugs. Old linen can be used as a slow-rotting mulch around the garden or added to the compost heap. Flax fiber is used in the making of paper.

Flax seeds are high in nutrients, especially magnesium, which most Western adults are deficient in. Seeds, which have a lightly spicy flavor, can be sprinkled over breakfast cereal, fresh fruit salad, and home-baked bread.

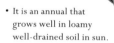

- It is an annual that grows well in loamy well-drained soil in sun.
- Its strength is in its tolerance of many climates enabling it to be grown from Egypt to India and China.
- Flax is well known for depleting the soil so it needs to be rotated as a field crop.
- Sow from seed either directly into prepared soil or into modules, planting out when roots have developed.
- Flax reaches a height of 3 to 4 feet (90–120 cm) in three months; it can then be harvested for its fibers. If grown for seeds, the plants need another month to develop and ripen.

Flax has various medicinal properties: antibacterial, antiviral, antioxidant, digestive, anti-inflammatory, autoimmune, and mood regulator. One of the essential omega-3 components is found in flax (alphalinolenic acid), which helps guard against many health issues, such as rheumatoid arthritis, cardiovascular disease, and blood pressure problems. It also helps lift the mood of those with depression. It is a good bulking agent to alleviate constipation. A tablespoonful of flax seed provides much-needed fiber in the diet.

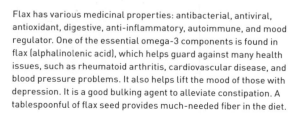

Lippia graveolens

Mexican Oregano

Mexican oregano, also called oregano cimarrón, *hierba dulce*, and redbrush lippia, is an aromatic shrub native to Mexico, Central America, and the southwestern United States. Being a member of the Verbenaceae or verbena family, the plant is not a close relation of the European *Origanum* species also referred to as oregano (see p.132). The common name refers to the oregano-like aroma and flavor of the plant, but in fact the flavor is hotter and much more intense. Dried leaves of the plant are exported from Mexico under the name of dried oregano. Thymol, an essential oil, is found in both *Origanum* and *Lippia* species, and this accounts for the similarities in taste between the two, and also for similarities in how they are used medicinally.

The Tarahumara Native Americans of northwest Mexico, who were dubbed the "barefoot running people," gather the leaves and cook them like spinach, often with beans. They also add them to toasted and ground corn in a dish called *pinole,* and to a cornmeal drink of the same name. The herb is part of Tarahumara natural medicine and is used to treat stomach problems and intestinal worms.

- Grow from seed with bottom heat and good light under glass, or from cuttings in late spring onward.
- This perennial plant is not frost hardy. It prefers light, well-drained soil in full sun
- A shrub or small tree, it attains a height of 3 to 9 feet (1–2 m)
- The small, verbena-like, yellow-white flowers appear in spring to late summer.

It provides nectar for bees, butterflies, and other nectar-feeding, pollinating insects.

The leaves can be applied externally as an emollient, or skin softener.

Antimicrobial, antiparasitic, and digestive, the herb is used in Mexico as a tea, called *guerro,* for stomach pains. It is also used to treat candidiasis (yeast infection), reduce fever, and regulate menstruation. The oil is used in aromatherapy.

The hot "oregano" taste of this plant is widely used in Mexican cuisine, especially with pork, chicken, fish, and pinto beans, and in bean soups. The pungent aroma and taste that it adds to salsa or meatballs with tomato sauce is authentically Mexican.

Lonicera periclymenum

Wild Honeysuckle

Common wild honeysuckle can be found growing in northern Europe (including the British Isles), western Asia, and North America. Its common name "honeysuckle" or "sweet suckle" refers to the practice of insects sucking nectar from the narrow end of the tubular-shaped flowers. The genus name was assigned by Swedish botanist Carl Linnaeus to honor the sixteenth-century German botanist Adam Lonicer, who wrote a popular herbal that contains unusual and curious information about plants. Honeysuckle is also known as woodbine and its summer flowers are intensely fragrant.

Periclymenum refers to the plant's climbing habit—"to twine around with tendrils"—and it is a useful garden plant for its ability to cover unsightly walls or buildings. In the language of flowers it symbolizes the uniting of two people in love and devotion. Honeysuckle has long been a symbol of fidelity and affection. Dante Gabriel Rossetti the English poet and Pre-Raphaelite painter wrote a poem entitled "The Honeysuckle" that is an erotic parable, which was published after his death in 1882.

According to folklore, allowing honeysuckle to grow around the entrance to your home protected against witches. Bringing honeysuckle flowers into the house was thought to bring money to those who lived there. The flowers were also crushed against the forehead to enhance psychic powers.

- Hardy deciduous perennial reaching 23 feet (7 m) in height and living up to fifty years.
- Fragrant yellow flowers appear from early summer to mid-fall and are followed by very red berries that are poisonous.
- Can be grown from seed in fall; germinates best with the seed left uncovered. Takes up to two years to germinate; cuttings from non-flowering shoots will be much quicker.
- Can also be layered.

Flowers are used to make teas, jams, jellies, and vinegars. They are often used to decorate cakes and desserts, and are especially good for making country wine.

Honeysuckle is used in cosmetics and perfumes. Honeysuckle flower lotion is good for cleansing and toning the skin.

Honeysuckle is an antiseptic, diuretic, and expectorant. The flowers have been used mainly in cough medicines. Historically, the leaves were used to relieve headaches, fevers, and the pains of rheumatism and arthritis.

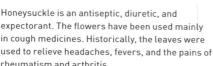

Malva sylvestris

Common Mallow

Common mallow is a spreading native herb found throughout Europe, growing wild along roadsides, wastelands, and construction sites. It sometimes grows abundantly on shorelines and has migrated to the northern states of North America.

English botanist and herbalist Nicholas Culpeper wrote in his *Complete Herbal* that all the mallows "are under Venus," referring to its use for affairs of the heart, although in the Middle Ages it had been used as an antidote to love potions and aphrodisiacs. The Romans ate the young shoots as a delicacy. The flowers have long been used in garlands during May Day celebrations and as a spring strewing herb. Mallow was once regarded as a universal cure-all in folk medicine. The plant has also been used as a natural dye.

Mallow leaves are rich in vitamins. Use only the young tender leaves and shoots in salads and for steaming as a vegetable. In Moroccan cuisine, the leaves are chopped, steamed, and then sautéed with olive oil, spices, olives, and preserved lemon to make a salad. Mallow seeds (known as cheeses) have a nutty flavor and can be eaten raw in salads.

Excessive usage may cause nausea and diarrhea. Do not use if pregnant or breast-feeding.

Common mallow produces a natural dye, which can yield cream, yellow, and green colors, depending on the mordant used in the cloth, wool, or fiber.

Common mallow is an anti-inflammatory, astringent, and purgative. Use as an infusion for coughs, bronchitis, sore throats, asthma, and emphysema. Also as a mouthwash for sore gums; gargle for sore throats and tonsillitis. Apply as a poultice to skin inflammation and insect bites.

- It is a hardy perennial that self-seeds freely. Produces rose purple to bluish mauve flowers from May to October.
- Sow seed in the spring.
- Likes a moisture-retentive soil in a sunny position; will also grow in dappled shade.
- If you want to crop the leaves, cut back in late June, water well, and feed for a second crop to use in September.
- Harvest flowers and dry quickly to retain color.

Mandrake

Native to the Himalayas and the southeastern Mediterranean region, mandrake has been used medicinally since biblical times, as well as by those interested in mysticism and magic. It was used in rituals for its hallucinogenic and narcotic properties and was an important element in exorcism, particularly in the Dark Ages.

The ancient Greeks first described its carrot-shaped root as being in the form of a human. Herbals later depicted the plant by drawings of a man with a beard and a woman with bushy hair. Many superstitions are attached to mandrake, one of which was that the roots screamed when pulled from the ground. As the person who pulled the roots was thought to be condemned to hell, animals were often used to extract them from the soil. Mandrake was said to grow under the gallows of murderers and, if dug from there, death would follow. It was often mixed with henbane and belladonna (see p. 37) in a witches' brew designed to kill off a rival. It was also made into amulets to ward off evil and bring good fortune. It has long been thought that the sponge given to Christ when he was crucified was dipped in a solution of mandrake, which worked as a sedative to ease the pain. Greek and Roman physicians used the plant as an early anesthetic, but it is rarely prescribed in modern herbalism.

- Hardy low growing perennial herb belonging to the nightshade (*Solanaceae*) family.
- Grows to a height of 4 inches (10 cm).
- Deep rooted so needs good loamy soil.
- Can be grown from seed of the ripe fruits or by division of the roots.
- Purple or white-greenish flowers appear from March to May; seeds ripen from July to August.

Mandrake is a sedative, purgative, and emetic. Once used as an analgesic and as a narcotic to induce hallucinations. Homeopathy uses potentized preparations of the leaf and root to treat headaches and other ailments.

The plant is highly poisonous (containing alkaloids hyoscyamine and scopolamine). Mandrake is rarely used today due to its high toxicity. It can cause death by respiratory paralysis.

Shrieks like mandrakes' torn out of the earth.

WILLIAM SHAKESPEARE,
ROMEO AND JULIET,
ACT IV SCENE 3

Marrubium vulgare ▲

White Horehound

Native to Mediterranean Europe, central Asia, central and northern Europe, and North and South America, white horehound is a perennial herbaceous plant found growing in wastelands, pastureland, and along roadsides. It has been used since ancient Egyptian times and up to the present day as a cough remedy. Egyptian priests called it the "seed of Horus" and it was known for its anti-magical powers, reversing curses, and for counteracting certain plant poisons.

It is a much-loved folk medicine that has naturalized in many places and is grown in herb and cottage gardens far and wide. A bitter herb and member of the mint family, its stems and leaves are covered in downy hairs giving it a silvery appearance that has resulted in its common name. It was once spelled "hoarhound" as in hoare frost. Native American tribes, such as the Cherokee, Hopi, and Navajo all used it for treating coughs. A cough lozenge containing the herb was sold by itinerant barrow boys in London. It has also been used to flavor horehound ale, a herbal ale very popular in Norfolk and Suffolk, and some liqueurs.

In the garden white horehound has been used as an insecticide against caterpillars and as a companion plant to tomatoes. It is noted for attracting wildlife.

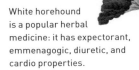

- An easy-to-grow hardy perennial that flowers in the second year.
- Succeeds in well-drained soil—flourishes in poor soil.
- Sow seed into plug trays or modules as the seed is small; germination takes two to three weeks.
- Take softwood cuttings from the new growth in early summer.
- Divide established clumps in spring, transfer into pots until rooting well, and plant back outside.
- Cut mature plants back after flowering.

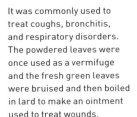

White horehound is a popular herbal medicine: it has expectorant, emmenagogic, diuretic, and cardio properties.

It was commonly used to treat coughs, bronchitis, and respiratory disorders. The powdered leaves were once used as a vermifuge and the fresh green leaves were bruised and then boiled in lard to make an ointment used to treat wounds.

> *It helpeth to expectorate tough phlegm from the chest.*
>
> NICHOLAS CULPEPER,
> *COMPLETE HERBAL* [1653]

Over a hundred
components have been
identified in the essential oil
of *Melaleuca alternifolia*.

CHERYLL WILLIAMS, *MEDICINAL PLANTS IN
AUSTRALIA, VOL. II* (2011)

Tea Tree

Melaleuca alternifolia is native to New South Wales and Queensland in Australia and is commonly found growing in the wild. Native Australians have long used this therapeutic herb to treat numerous ailments. Revered as a powerful bush remedy, the leaves were crushed and used as a body rub and even the water collected from underneath the tree would be used as a healthy wash. It is thought that the name comes from the observation that the shed leaves turn water tea-colored.

The essential oil is the most effective and prized product of this coniferous tree and it is produced on a commercial scale by steam distillation of the leaves. Use of the oil remained within Australia and was known as a bush remedy until the twentieth century. In 1922 Sydney chemist Arthur Penfold began to distill the oil and published a paper extolling its ability to treat bacterial and fungal infections. During World War II, Australian forces used it as an antiseptic to dress wounds, which caused it to be in short supply at this time. Production of the oil declined when synthetic substitutes were deemed cheaper. In the 1990s natural tea tree was reintroduced when it was found to treat certain strains of *Staphylococcus* more successfully than antibiotics. It is now widely used in complementary therapies, surgery, burn care, dental care, and by the cosmetic industry.

- Found growing in the wild along streams, water courses, and swampy flats; also on the coast.
- Suits a woodland garden near the sunniest edge.
- Propagation from seed is quite difficult; sown seed needs to be kept moist, only slightly covered. Needs good light levels and warmth; should be at a constant temperature. Flowers are hermaphrodite and appear in June. Not frost hardy.
- Oil can be used fungicidally in a diluted spray on other plants.

Tea tree is a gentle natural remedy with a powerful kick: it has antibacterial, antiseptic, and antifungal properties. Oil is obtained from the leaves and twigs. The essential oil can be diluted to treat stings, burns, wounds, and skin infections. It also stimulates the immune system. It can be used externally to treat thrush and vaginal infections. Apply undiluted essential oil to verrucas and warts. For all other uses, dilute tea tree oil in almond oil.

Added to water, tea tree oil can be used as an all-purpose spray cleaner. In low diluted concentrations (below 1 percent), it can be safely used to treat flea infestations in dogs and cats.

Not to be taken internally as it causes severe vomiting, diarrhea, and gastric irritation. Tea tree oil can cause skin irritation if used topically in high concentrations.

Tea tree is used in low concentrations in many cosmetics. Its germicidal properties make it a useful addition to dental products, such as mouthwashes, deodorants, and soaps.

Described as "a medicine cabinet in a bottle," one of the many wonders of tea tree is its use as a tissue regenerator. In aromatherapy, it is used to alleviate mental stress.

Sweet Melilot

Native throughout Europe, Asia, and the United States, sweet melilot is also found growing wild on waste grounds and roadsides in Britain. It naturalized there when it was extensively grown for cattle fodder in the eighteenth century.

Also commonly known as sweet clover, sweet melilot has been used since ancient Egyptian times for both culinary and medicinal purposes. Medicinally, the Egyptians used it to treat earache and worms. The Greeks used melilot poultice to draw out poison; they also used it dried for its sweet scent, calling it the "honey lotus" from the Greek *meli* (honey) and "lotus," referring to its sweet-scented flowers. It is an excellent nectar plant for honeybees and is especially good for wild herbal habitats.

The whole plant contains courmarin, which gives it a strong sweet almond scent similar to sweet woodruff (see p. 75), hence its latter day use for strewing. It was also once used in snuff and as a moth repellent.

! Not to be taken unless under the supervision of a qualified practitioner. If it is allowed to ferment while drying it produces dicoumarol, an anticoagulant that is used in rat poison. If not dried correctly, it is also toxic to pets and livestock due to a mold that forms on the herb.

 Only fresh sweet melilot leaves should be used because they become toxic when dried. Use in rabbit dishes, stews, and stuffing. Historically, it was used as a cheese flavoring. The fresh flowers can be used in tisanes, salads, and for cake decoration.

 Sweet melilot is an aromatic, calmative, expectorant, antispasmodic, and antibiotic. It is used by herbalists to treat thrombosis, varicose veins, and phlebitis. Historically, it was used to alleviate swellings and bruises. A tea made from the fresh plant aids digestion.

- It is a member of the legume (pea) family.
- Annual or biennial, depending on when the seed is sown.
- Grow from seed in early spring; will self-sow.
- Harvest the flowering tops and leaves for drying when all the flower spikes are open.
- Sown and used as a green manure that enriches the soil and provides good organic matter.

Melissa officinalis

Lemon Balm

Lemon balm is a popular perennial herb that originated in the Middle East and has naturalized in the Mediterranean and central Europe. It found its way to Britain via the Romans, who embraced it for its medicinal properties and it later became an important herb in monastic gardens. It is known as the "bee herb"—the generic name is derived from the Greek word *melissa* for honeybee—because its flowers are so attractive to them.

Ancient medics extolled the ability of the herb to "make the heart merry" and revive spirits. English writer and gardener John Evelyn described the leaves steeped in wine as having a "cordial and exhilarating effect," whereas the herbalist John Gerard commented that "it driveth away melancholy and sadness."

The Greeks used lemon balm leaves laid on bee hives to keep bees from leaving and to guide them back to the hive. Lemon balm is often planted at the four corners of an orchard to attract bees to pollinate the fruit trees. Beekeepers drink lemon balm tea before dealing with a "touchy" swarm because it has a soothing effect on both keeper and bees.

- Easy-to-grow herb, prefers fairly rich, moist-retentive soil in a sunny position. Grows vigorously and will spread into other planting.
- Sow seed in the spring in plug trays; do not over-water; plant outside when fully rooted.
- Cuttings from new growth.
- Divide plants in early spring and fall.
- Harvest leaves before the flowers appear. Best used fresh; dried loses its potency quickly.

Lemon balm has various medicinal properties as a calmative, antidepressant, and antiviral. A crushed leaf applied to an impending cold sore will slow its progress. It is also used for insect bites. Use in a tea to alleviate nervous tension and headaches, fatigue, nervous exhaustion, and upset stomach. Lemon balm lifts the mood and helps with depression and the oil is used in aromatherapy for this.

Use fresh lemon balm leaves in wines, cordials, and cold drinks. They are also good in fish and poultry dishes; add to sauces, marinades, jams, jellies, fruit salads, and sorbets.

 Fresh lemon balm leaves can be used to polish wooden furniture. The oil in the leaves cleans and shines the wood, as well as leaving a pleasant lemony scent.

But woe unto you, Pharisees!
for ye tithe mint and rue and all
manner of herbs, and pass over
judgment and the love of God.

THE BIBLE, LUKE 11:42

Mentha spp.

Mint

Many varieties of mint are in cultivation today, including spearmint (*Mentha spicata*), applemint (*M. suaveolens*), and peppermint (*M. x piperita*), as well as numerous cultivars, all with individual flavors, scents, and textures. The herb, believed to have originated in the East, was introduced to Europe via North Africa, and people in Arabic countries have long taken mint tea as a social drink and for virility.

Spearmint was considered an important herb by the ancient Egyptians, Greeks, and Romans, and is often referred to by Dioscorides, Hippocrates, and Pliny as a valuable medicine, strewing herb, food enhancer, and aromatic. The Greeks were known to use it as a cleanser in the home, using it to scour tables before meals; they also added mint to their bathwater. The Romans used it as a food flavoring and as an appetite stimulant, and it was as such that they introduced it to Britain. Records as early as the third century CE exist of mint sauce being made in Britain as an accompaniment to meats, and by the sixth century mint was being used to clean teeth.

In folklore it was suggested that mint would bring luck and prosperity if a few leaves are rubbed into a purse.

- Mint is a perennial aromatic plant with a creeping root system that enables the plant to spread rapidly, depending on the species.
- Grow from tip cuttings or division of the runners.
- Because mints hybridize readily, they cannot be guaranteed to come true from seed if several varieties are grown in a bed together.
- Mint grows well in containers; it is best to devote one to mint alone—it crowds out other herbs.

Mint repels insect pests, especially fleas and beetles, so add it to the bedding of dogs and cats. Mice and rats dislike mint, too, so soak a cloth in peppermint essence and put it into the entrance of mice and rat runs to drive the rodents away.

Use spearmint as mint sauce with lamb, or to flavor jellies, yogurt, crème fraîche, or sour cream. Peppermint goes well with chocolate, used in butter icing, mousse, or cream.

Chilled, mint-infused water makes a refreshing facial tonic. Fresh mint rubbed on the face, neck, and hands will repel biting insects.

Peppermint is the species most used for its healing properties. Make an infusion for gastric problems, indigestion, irritable bowel syndrome, and colic. It promotes sweating, so use it for colds and flu, as a gargle for sore throats, and as an inhalant for blocked nose, sinuses, and catarrh.

In Greek mythology, the beautiful water nymph Minthe was pursued by Pluto, God of the Underworld. She was to feel the wrath of his wife, Persephone, who turned her into a plant that would be trodden under foot. Unable to reverse the deed, Pluto caused mint to release a beautiful fragrance.

Do not use mint as an inhalant if you have asthma, and do not give it to young children. Avoid constant use during pregnancy.

MINT | 117

Bergamot

Bergamot, also known as scarlet bee balm and Indian plume, is a native of North America and is famous as the source of Oswego tea, which is named after a district near Lake Ontario where the herb grew wild in abundance. Native Americans from the area made tea from both *Monarda didyma* and *M. fistulosa*, and for a while this replaced tea from India after the Boston Tea Party of 1773.

The genus name *Monarda* is derived from that of Nicholás Monardes of Seville who, in 1569, published an herbal on the flora of America, drawing on his insights as a medical botanist.

The common names of bergamot and red bergamot refer to the scent of the crushed leaves, which resembles the fragrance of the small, bitter bergamot orange from Italy, whose oil is added to tea to make the Earl Grey blend, originally formulated as an imitation of expensive Chinese teas. The bergamot orange is also used in perfumery. *M. didyma* is also known as bee balm because it is very attractive to bees, and also hummingbirds, which are drawn to the long, tubelike, nectar-rich flowers.

- *M. didyma* is the most aromatic species of its genus.
- It prefers rich, moisture-retentive soil with plenty of nutrients.
- Semi-shade is preferred, but it will tolerate sun as long as the soil remains moist.
- The plant can be grown from seed but is more reliably cultivated from softwood cuttings in early summer.
- Propagation is also possible by root division in spring. Divide mature plants every three years to keep the clumps healthy and vigorous.
- It is prone to mildew.

The plant is an antiseptic, anti-emetic, digestive, expectorant, and sedative. Make an infusion for indigestion, nausea, and flatulence. Its rich thymol is good for sore throats in a gargle. The oil is used in aromatherapy for depression and insomnia.

Bergamot and Oswego tea are used for premenstrual syndrome and should be avoided during pregnancy.

The edible flowers are used in fruit salads and fruit cups, and as cake decoration. Leaves may be added to leaf salads, jams, and jellies.

For the elderly, bergamot tea is the most comforting taken last thing at night.

MARGARET ROBERTS, *EDIBLE AND MEDICINAL FLOWERS* (2000)

Curry Leaf

Curry leaf is a tree native to India, Pakistan, and Sri Lanka, and is now cultivated in Thailand, Malaysia, Indonesia, China, Australia, and other tropical countries. The fresh leaves are used extensively in curries in the southern regions of India and Sri Lanka. Dried leaves can be purchased in specialist shops in the West but they are of little culinary worth because the drying process causes the leaves to lose most of their aroma and flavor.

Largely due to the influence of migrant Indians, curry leaves are now added to numerous dishes of many different cultural cuisines. Away from its homelands, curry leaf is especially favored with other spices in fish curries prepared by ethnic groups in Malaysia, Singapore, and the Thai islands. Curry leaves may be regarded as the Indian and Asian equivalent of the bay used in Western cuisine and is often removed from a dish before serving.

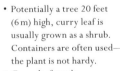

Curry leaf has antibacterial, antifungal, and anti-inflammatory properties and is used in Ayurvedic medicine as a treatment for many different types of infection. It is also antidiabetic and antioxidant and is used to treat diabetes and high cholesterol, and to protect the liver. In other medical disciplines, curry leaves are used to treat peptic ulcers, dysentery, and poor eyesight—the latter taking advantage of its high content of vitamin A.

- Potentially a tree 20 feet (6 m) high, curry leaf is usually grown as a shrub. Containers are often used—the plant is not hardy.
- Curry leaf needs temperatures of at least 59–64°F (15–18°C).
- Prefers moist, humus-rich, well-drained soil, in sun or partial shade.
- It can be raised from seed in spring, kept under glass at a constant temperature, and also by semi-ripe cuttings in summer.
- Leaves may be harvested all year round.

Whole leaves are infused in coconut-milk-based curries with fish and vegetables. The leaves are also finely chopped or minced for adding to curries, marinades, and omelettes.

The oil extracted from the seeds of the fully ripe, black, shiny fruits strengthens hair roots, promotes healthy hair growth, and slows the loss of pigment and encroaching grayness in a head of hair.

Nutmeg and Mace

Nutmeg and mace are two different but closely related spices derived from the fruit of the nutmeg tree, a native of the volcanic soils of the Moluccas and other islands of the East Indies. Both are known to have been imported in quantity by Arab countries and India as early as the sixth century CE, and there are reference to their being in use in Europe by the late twelfth century. The spices were then valued more for their medicinal value than as a flavoring.

When the Portuguese took possession of the Moluccas in the early sixteenth century, they promoted both spices for their tonic effect. Their monopoly in trade was later seized and continued, first by the Dutch and then the English, until the beginning of the nineteenth century. Famously, the Dutch traded the island of Manhattan for control of the last nutmeg-producing island then still under English dominion.

- *Myristica fragrans* is a tender plant that needs heat and humidity; in northern latitudes, a tropical glasshouse is essential.
- The plant can be grown from woody cuttings, placed in moist, humus-rich soil.
- Even in optimal conditions, plants are likely to wait nine or ten years before producing any fruit.

Oil pressed from nutmeg kernels is used in the manufacture of soaps and perfumes, and in candle making.

Nutmeg is an astringent, anti-inflammatory, digestive, sedative, and aromatic. Used in small doses only, it can reduce flatulence, aid digestion, improve appetite, and combat nausea.

Blades of mace, the outer casing or aril of the nutmeg seed, are used to flavor milk puddings, rich fruit cakes, cheese sauces, and béchamel (white) sauce. The kernels are often finely grated onto finished dishes such as baked fish and rice pudding. Used sparingly, nutmeg adds pungent flavor to cakes, cookies, preserves, and mulled wine.

Use nutmeg very sparingly; excessive quantities can cause hallucinations, heart problems, and epileptic convulsions.

Myrrhis odorata

Sweet Cicely

Sweet cicely is a native of Europe, parts of Russia, and Asia, where it is found growing wild in hedgerows, roadsides, and on high ground. It was once cultivated in pots as a "pot shrub." The genus name *Myrrhis* is derived from the Greek word meaning "smelling of myrrh," while the specific name *odorata* is a reference to the plant's pleasant fragrance. Sweet cicely is the sole species of its genus.

Historically, a decoction of the plant's fragrant antiseptic root was used to treat bites from dogs and snakes, and the plant also yielded an ointment for treating gout and leg ulcers. The roots were praised by herbalists, who candied or boiled them as a warming and comforting protection against the plague.

The plant should not be mistaken for American sweet cicely, *Osmorhiza longistylis*; with similar white umbel flowers, and used by Native Americans for medicinal purposes, that is clearly a different plant altogether.

Use leaves fresh, immediately after picking to preserve their potency. They make a bath for sore eyes and an infusion for coughs and minor digestive problems. Roots are best collected and dried in the fall. Use in a poultice for boils and deep wounds, and in a syrup for coughs.

Sweet cicely is loved by bees, and it produces good honey. The seeds used to be crushed into a paste to make an aromatic polish for oak furniture and flooring.

- In early fall, sow seeds individually in plugs, cover with glass, and place outside in winter. The seeds require the cold to germinate.
- Covering the seeds with glass is necessary because rodents will otherwise devour them voraciously.
- Take root cuttings in spring or fall.
- Sweet cicely dies down in fall and reappears in spring; leaves may be harvested at any time.

Sweet cicely serves as a natural substitute for sugar when cooking tart fruits, rhubarb, and gooseberries, and is especially useful for diabetics and people who need to avoid sugar. The flavor is like mild aniseed. The roots may be boiled and eaten in a salad with a vinaigrette dressing, and leaves added to salads, fruit salads, dressings, soups, and drinks. Highest in oil content are the seeds, which may be eaten whole or added to fruit pies, as cloves are.

Myrtle

Myrtle is a bushy evergreen shrub native to southern Europe and Southeast Asia. Half-hardy in temperate climates, it is cultivated as a garden species throughout Europe, where it is often grown as a specimen plant. It grows to a maximum height of about 10 feet (3 m).

Myrtle has long been of interest to humankind and is frequently mentioned in the Old Testament and the writings of ancient Greek and Roman poets. The genus name is derived from *myrtos*, the name by which it was known. During certain military processions in ancient Rome, commanders were crowned with wreaths of myrtle.

The herb is dedicated to the goddess Venus and is considered an aphrodisiac. According to legend, Venus transformed Myrrh, one of her priestesses, into a myrtle tree to protect her from overzealous suitors. From ancient times to the present day, myrtle has been incorporated into wedding bouquets and head crowns as a signifier of love, constancy, fidelity. A sprig of myrtle from Queen Victoria's wedding bouquet was planted, and sprigs from the tree are used in royal wedding bouquets.

In the Middle East, powdered myrtle leaves were used for dusting babies before they were wrapped in swaddling.

- The kidney-shaped seeds are sown in spring.
- Propagate by semi-ripe cuttings or layering in late summer.
- Myrtle likes a sunny position, preferably against a sheltering wall.
- It requires protection against heavy frosts, especially when young. Large plants are only affected below 14°F (-10°C).
- Harvest leaves throughout the year, and collect the fruits when ripe for drying.

An oil made from the leaves and flowers is used in soaps and skin-care products. All of the plant is aromatic and suitable for potpourri.

It was roses, roses . . . with myrtle mixed in my path like mad . . .

ROBERT BROWNING, "THE PATRIOT" (1855)

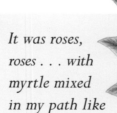

Add fresh flower buds to salads after removing the green part. Leaves may be added to roasting pork, wild boar, and lamb for the last ten minutes of cooking to impart a delicate aroma and flavor. Fresh, ripe berries, or alternatively dried ones, may be used sparingly in marinades for meats.

Myrtle leaves have antibiotic, astringent, and antiseptic properties, and were once used to treat vaginal problems. They were also used externally as a poultice to treat hemorrhoids.

Nepeta cataria

Catmint, Catnip

Also known as catnip, *Nepeta cataria* is a true member of the mint family. With aromatic leaves and white flowers, it is not as decorative as blue-flowered catmint species, but it is the herb that historically has been used to treat a number of ailments. It is a hardy perennial, native to Europe and Asia, and is now naturalized in North America and South Africa.

The common name refers partly to its allure to cats, for whom it appears to act as an aphrodisiac. It is believed that the aroma exuded by the herb when withering or bruised is similar to the pheromones of cats of the opposite sex. Seemingly intoxicated by the bruised parts, cats can actually destroy them by rolling in them and eating them. In the experimental hippie era, some people dried and smoked catmint leaves for their reputed mildly hallucinogenic effect.

Catmint leaves have been used to make sauces and dressings, and may be added in small amounts to salads, as may the flowers. The herb is infused as a hot tea, sometimes mixed with lemon balm (see p. 115), or iced as a refreshing tonic. It is also used as a rub for meats.

Catmint is an anti-inflammatory, digestive, and relaxant. Rubbing crushed leaves onto the skin will keep away biting insects and relieve swellings. An infusion of the leaves makes a soothing and relaxing treatment for insomnia, nervous excitability, indigestion, colic, and flatulence.

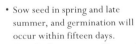

- Sow seed in spring and late summer, and germination will occur within fifteen days.
- Can also be propagated by cuttings.
- Division should be done out of the "scent range" of cats because they have been known to destroy freshly divided plants.
- Catmint likes a position in full sun, in well-drained soil; it will rot in a wet situation.
- Cut back hard after flowering to promote health and vitality in mature plants.

Pregnant women should avoid taking this herb internally because it adversely affects the uterus.

Catmint repels fleas, aphids, flea beetles, ants, mice, rats, and voles. Plant it by salad crops to deter flea beetles. Bees love the plant, however.

In Italy, where sweet basil is called "kiss me Nicholas" (bacia-nicola), it is thought to attract husbands to wives.

"BASIL: AN HERB SOCIETY OF AMERICA GUIDE," ON THE HERBSOCIETY WEBSITE

Basil

Native to India and the Middle East, basil is a highly aromatic herb that was brought to the Mediterranean by spice traders. The ancient Egyptians passed on their knowledge of basil to Arabia, Greece, and Rome, with Pliny and Dioscorides extolling the virtues of the herb in the first century CE. The Greek name of *basileus*, meaning "king," suggests the perceived importance of the herb. The ancient Greeks believed scorpions bred under pots of basil, and the Romans saw the herb as a symbol of hatred. These negative associations were turned on their heads in later times in Italy when basil became a token of love. There is no record of basil in Britain until the sixteenth century, but from there it was introduced to America and Australia by settlers.

Numerous superstitions were associated with basil. It was thought that cursing while sowing the seed caused the plants to grow with more vigor; carrying the herb was believed to bring wealth; and a pot of basil presented as a house-warming gift was expected to bring good luck. Basil was considered purifying and able to ward off evil.

Basil is revered in India, especially *Ocimum sanctum*, or holy basil, which is sacred to the gods Krishna and Vishnu. Thought to protect against evil, it is planted around Hindu temples. It is also used in burial rites, with holy basil being placed in the casket before burial.

- Sow basil seed from early spring to midsummer.
- Sow into plug trays, not seed trays, because it hates to have its roots disturbed.
- As a tender annual herb, basil needs constant warmth and light to thrive, and is best grown under glass or in pots in the sunniest part of the garden. It can damp off with too much temperature fluctuation.
- Pinch out young plants to stop them flowering.
- Can be sown directly into the ground, but only when there is no threat of frost.

Having a lamp of basil oil burning while studying is said to help concentration and uplift the mood, while a massage with basil oil is performed to alleviate tiredness and depression.

Basil leaves may be ground with hibiscus leaves and flowers to make a combined shampoo and conditioner. Basil leaves simply boiled in water make an effective, antiseptic face cleanser.

Basil was widely associated with rituals of initiation, especially during the sabbath of Candlemas—the Presentation of Jesus at the Temple—which was thought of as a time of renewal.

Basil is best eaten fresh, and should be added at the very last minute to a cooked dish because the flavor soon deteriorates. It is used in pesto and tomato dishes, especially pasta and pizza; with chicken, zucchini, and peppers; and stir-fried with broccoli, garlic, and chili. It is preserved in oil in Italy, and in salt in India.

The leaves are used for mouth ulcers, insect bites, and skin eruptions, and in a gargle for sore throats and gums, ulcers, and oral thrush. An infusion can treat colds and flu, fevers, digestive problems, and constipation. Basil also counters nervous headaches and migraines, bacterial infections, and intestinal parasites.

In a dark night . . . this plant, when in full flower, can be seen at a great distance, having a bright white appearance, which probably may arise from some phosphoric properties of the flowers.

FREDERICK TRAUGOTT PURSH, QUOTED IN *KING'S AMERICAN DISPENSATORY* (1898)

Evening Primrose

Evening primrose is a native plant of North America, where it is found growing on dry wastelands and is regarded as a weed. Elsewhere in the world it is considered a valuable medicinal plant as well as a beneficial and attractive garden plant.

The common name derives from a belief that the flowers open between six and seven o'clock in the evening; in fact, they may open at any time of the day, depending on the weather. The opened flowers pour forth a sweet fragrance that attracts the moths that pollinate the plant; this phenomenon explains the folk name of "moth moonflower." A third folk name, "evening star," derives from the fact that at night the petals emit phosphorescent light, which is thought to attract moths.

Native American tribes such as the Iroquois created a wash from the roots to treat hemorrhoids. The Ojibwa soaked the whole plant and applied it to bruises. The Lakota burned the seeds as an incense as well as using them to obtain a fragrance.

In addition to medicinal and other applications, the Cherokee used the leaves as a green vegetable, and the roots in the same way as potato. When the evening primrose was introduced to Western gardens in the eighteenth century, it was primarily as the source of a root vegetable known as yellow lamb's lettuce, or German rampion.

- Seed can be planted in situ or grown in seed trays, starting in late spring.
- Leave the seed uncovered because it needs light for germination.
- Once planted, this hardy biennial herb will inevitably self-seed quite prolifically.
- The plant likes a well-drained, dry, sunny spot. Keep the site weed-free
- Harvest leaves before the plant flowers and use fresh.
- Collect the seeds for pressing as soon as they are ripe.
- Dig up the roots in the fall of the second year.

The oil extracted from evening primrose oil seed has proved especially helpful to women, being effective in alleviating pain, breast tenderness, and other symptoms of premenstrual syndrome (PMS), as well as symptoms of menopause such as hot flashes. It is used to treat the inflammation associated with endometriosis. It helps women with fibrocystic breasts by increasing their absorption of dietary iodine, which can be brought very low by the condition.

Evening primrose oil is used in a number of cosmetics and skin-care products, especially for mature and sensitive skins. Use the fresh plant for a facial steam.

The fresh young leaves, flower buds, and petals of the potted herb may be used in salads or pickled. The parsnip-like roots can be lifted at the end of their second season for steaming or use in soups or stir-fries.

The oil is used commercially for its phenylalanine, omega-6, and gamma-linolenic acid content. In addition to women's disorders, it can treat chronic headache, rheumatoid arthritis, diabetic neuropathy, anxiety, insomnia, eczema, dermatitis, and other disorders of the skin.

Do not use during pregnancy unless it is required at term to initiate or shorten labor.

Olea europaea

Olive

The olive tree is thought to have originated in Asia Minor and has been cultivated from ancient times for its life-giving fruit and oil or "liquid gold," as Homer called it.

Olive trees were once considered so sacred that if anyone cut one down, they would certainly be sent into exile, or even condemned to death. In early writings, the trees were seen as symbolizing happiness and peace; in the Old Testament, for example, the dove returns to Noah in the Ark carrying a sprig of olive, a welcome sign that the flood was abating. The oil was used as fuel in the sacred lamps of Greek temples, and winners in the early Olympic Games were crowned with olive leaves. The Greeks also anointed their dead with olive oil.

Thomas Jefferson is credited with the introduction of the olive to the United States. He wrote that "the olive tree is the richest gift of Heaven," and that bringing it to America was one of his important achievements.

Today there are around 800 million olive trees growing around the world. Spain, Italy, Greece, and Turkey are the main producers, but olives also grow in quantity in southern France and numerous other Mediterranean countries, North Africa, and the Middle East. Each producer has its own unique varieties, which are carefully graded for quality and either used for oil or reserved for the table.

- Propagate olive plants by semi-ripe cuttings in summer.
- Requires a well-drained position. In cool climates, a wet winter can kill the tree.
- The frost hardy, long-lived evergreen tree will start bearing fruit in its seventh year.
- The leaves may be picked at any time; the fruit is harvested both unripe and ripe for eating, and ripe if it is to be pressed for oil.
- There are records of olive trees that have lived for up to 1,500 years.

Fresh olives taste very bitter and must be pickled in brine before they can be eaten. Both pickled olives and olive oil are central to Mediterranean cuisine, one that has been shown to be life-prolonging and one of the healthiest in the world. Regular consumption of good-quality olives and oil, in tapenades, sauces, marinades, and dressings, is thought to prevent circulatory disease. Olive oil has long been in use as a preservative of foods due to its resistance to oxidation.

The olive is astringent, antibacterial, antiviral, and antioxidant in its effects. Historically, the leaves were made into a tea to treat feverish illnesses such as malaria, and to make poultices for skin infections and inflammation. The tea is also said to strengthen the cardiovascular system. Consumption of olive oil is thought to help reduce digestive hyper-acidity. Taken with lemon juice, olive oil is used to treat gallstones. The oil is also mildly laxative.

Olive oil contains antioxidants, including vitamins A and E, and is used for soaps, shampoos, and skin-care products. Some women use olive oil as a gentle make-up remover.

Roman women were among the first to use olive oil to treat the skin pigmentation and scarring of stretch marks, caused by pregnancy or weight changes. The oil is still popularly used today.

The olive-grove of Academe . . .
where the Attic bird
Trills her thick-warbled notes
the summer long.

JOHN MILTON, "PARADISE REGAINED" (1671)

*According to Roman legend,
Venus, goddess of love, gave
the plant its scent to remind
mortals of her beauty.*

"OREGANO AND MARJORAM," ON THE HERB
SOCIETY OF AMERICA WEBSITE

Sweet Marjoram

Sweet marjoram is the most aromatic of all the *Origanum* genus, with a warmth and intensity that gives a deep, rich flavor to cooking. The plant is a native of the Mediterranean, North Africa, the Middle East, and parts of India, and still can be found growing wild in these regions. It is now cultivated in many parts of the world, including the United States, and is often grown in pots or beds with other herbs.

The ancient Greeks saw marjoram as a symbol of happiness, harmony, and peace, and therefore planted it on graves to help the dead attain eternally undisturbed rest. The Greeks had more light-hearted beliefs associated with the herb, too—such as that if you slept with marjoram under your head, you would dream of your future love. The herb was also used to make bridal garlands and wreaths.

In the Middle Ages the leaves were chewed to alleviate toothache and aid the "passing of indigestion." The seventeenth-century herbalist Nicholas Culpeper observed that:"It is an herb of Mercury, and under Aries, and therefore is an excellent remedy for the brain and other parts of the body and mind, under the dominion of the same planet." Culpeper saw it as "warming and comfortable in cold diseases of the head, stomach, sinews and other parts, taken inwardly or outwardly applied."

- In the spring, sow seed, uncovered, in plug trays, with a bottom heat of 60°F (15°C). Germination is unpredictable, from sporadic to 100 percent.
- Do not over-water the seedlings. Prick out if over-crowded then transfer to grow on, or plant directly into prepared ground.
- The half-hardy perennial (an annual in cooler climates) likes a sunny, well-drained position.
- The whole of the plant can be harvested.

Sweet marjoram is used to flavor many Mediterranean dishes, being especially good with artichoke hearts, asparagus, and mushrooms. It also features in salads, and tomato- or vegetable-based soups and sauces. It is mixed into both meat for sausages and stuffings for meat dishes. With thyme, basil, tarragon, and chervil, marjoram is a component of the classic herb mixture called *bouquet garni*. The herb is also used to flavor marinades, oils, vinegars, and savory jellies.

Marjoram was popular as a sweet-smelling strewing herb, and its disinfectant properties helped to ward off disease and infection in the home. Placed in sachets, it was effective in keeping linens free of insects.

An infusion of marjoram helps to settle nervousness and prevent insomnia. It can aid the digestion and ease gastrointestinal disorders. Bath preparations can alleviate rheumatism.

Marjoram produces small flowers from knot-like buds—hence the common name "knotted marjoram." These help to distinguish it from the culinary herb oregano.

Marjoram tea is an anaphrodisiac—it calms hypersexuality and unwanted stirrings of the libido in both men and women. Religious orders sometimes used it as an aid to sexual abstinence.

Oregano

Oregano is a native plant of Europe, Iran, the Middle East, and the Himalayas. Its common name is wild marjoram and it is often confused with its cousin, sweet marjoram (*Origanum majorana,* see p. 130), with which it sometimes grows. Sheep and goats graze happily on both herbs but cattle disregard them. A higher proportion of thymol makes this herb taste more strongly of thyme than sweet marjoram.

The ancient Egyptians regarded oregano as being able to heal, disinfect, and preserve. The Greeks named the herb *oros ganos,* or "joy of the mountain" and they crowned bridal couples with oregano and other *Origanum* species. The Greeks would harness the calmative and therapeutic powers of its oil by massaging the whole of the head with it. The Romans, who introduced the plant to Britain, saw it as a plant of happiness and marital harmony. Eventually oregano reached North America, carried there by European settlers.

In the past, the flowering tops were processed to obtain a dye that turned woolen cloth purple, and linen a reddish brown, although in neither case was the dye color-fast. The leaves, meanwhile, were used in bunches to polish and scent good-quality furniture.

Before the introduction of hops, oregano was among the herbs used to flavor ale. The flowering tops would also preserve the beer.

- In the spring, sow the very fine seed in plug trays. Leave uncovered, with a bottom heat of 60°F (15°C) or warmer to stimulate germination.
- Propagate also from new-growth cuttings and from division in spring.
- Do not over-water; plant outside when the roots fill the plug.
- Site in well-drained, dry, alkaline, nutrient-rich soil, in full sun to produce a stronger flavor.
- Cut back mature plants after flowering.

Oregano is antifungal, anti-inflammatory, antioxidant, digestive, and relaxant in its effects. Add the leaves, flowering tops, or oil to a hot bath to soothe tense muscles and sore joints. A mouthwash may be made for sore gums and mouth ulcers. Drinking an infusion will ease a nervous headache and anxiety, and cure insomnia by promoting calm, restful sleep. An infusion can alleviate menstrual pain and regulate menstruation. It will also soothe a cough and help to clear congested sinuses.

Oregano is a major flavoring in Italian, Greek, and Mexican cuisine. It is an ingredient of the classic Sicilian Salmoriglio sauce, used for fish, lamb, or venison, and is delicious with pumpkin and omelet, as rubs for breads or meats, and as a pizza seasoning. The French use it with salsify, cucumber, and carrots, and it enhances zucchini, eggplant, and beans. Add oregano at the end of the cooking because it becomes bitter if cooked for too long.

In folklore, one of the more unusual recommendations is that wild oregano, marjoram, or thyme be laid by the side of milk in a dairy to "prevent it from turning if there should be thunder."

Like many herbs, oregano should be avoided during pregnancy. Consumption of large amounts can cause miscarriage, and how safe it is at much lower levels of ingestion is not known.

Tomatoes and oregano make it Italian; wine and tarragon make it French.

ALICE MAY BROCK, FOUNDER OF
ALICE'S RESTAURANT

Oxalis acetosella

Wood Sorrel

As its common name suggests, in the wild this plant is mainly found growing in woodlands and shady places. It is native to Europe, northern and central Asia, and Japan; it is naturalized with other *Oxalis* species in North America.

The genus name *Oxalis* is from the Greek for "sour" or "acid," while *acetosella* means "vinegar" or "salts." Wood sorrel is known to have been cultivated as a much-appreciated sauce herb from the fourteenth century, but it lost some popularity after the introduction of French sorrel (*Rumex scutatus*).

In Europe, apothecaries used to mix wood sorrel with copious amounts of sugar and some orange peel to make *conserva liguloe*, a cooling acid drink devised as a remedy for scurvy. By the time of the reign of Henry VIII it had become a highly regarded herb, mainly used as a pleasantly bitter addition to salads.

If direct sunlight falls on the leaves, they shrink down to form a three-sided pyramid to prevent evaporation from the pores. At night the leaflets fold in half along the midrib and lie almost side by side to "sleep." This is thought to be a natural protection against storms and excessive dew.

- Sow seeds in a cold frame as soon as they are ripe. Prick out and transfer the seedlings when they are large enough—plant out in spring.
- Roots divided from plants in spring and placed in a moist, shady border will soon grow into new plants.
- Large divisions may be planted directly into the ground; small ones fare better after a time in the cold frame.
- Keep the site clear of weeds or the wood sorrel is easily overwhelmed.

The leaves and flowers are rich in vitamin C. Use in salads, and in place of French sorrel for green sauces to accompany fish.

The juice of wood sorrel leaves (which turns red when clarified) was once used diluted in a gargle for sore throats or mouth ulcers, or to staunch bleeding gums. A sponge or lint soaked in the juice helped to reduce swelling, inflammation, or heat caused by infection.

The plant is slightly toxic due to its high content of oxalic acid. It also interferes with absorption of some of the trace minerals in food, including calcium, with which it seems to bind.

People with kidney disease, stones, gout, or rheumatoid arthritis should not consume wood sorrel.

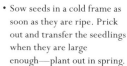

Passiflora incarnata

Passion Flower

A highly ornamental climber or vine, *Passiflora incarnata* is just one of several species known as passion flower. It is a native of Mexico, Central America, and the southern United States, where it is a common wild flower. It is the most common edible form of passion flower, producing yellow, egg-shaped fruits from beautiful, complex flowers that are white with a lavender or purple crown. The vine is one of the largest of its genus, grows rapidly, and has a pineapple-to-orange scent. The species name, *incarnata*, means "in the flesh" and refers to the texture of the fruit.

The passion flower was discovered in Peru in 1569 by the Spanish physician and botanist Nicholás Monardes. Much later it was to become a popular plant in nineteenth-century Britain, where plant breeders realized that the entire genus responded well to hybridization—consequently, hundreds of different forms were bred and distributed around the British Empire.

Native North Americans were familiar with the plant and used it to heal cuts and bruises, and as a tonic; they also favored the fruit as a food. The Cherokee called it *ocoee*, and the Ocoee River and its valley in the southern Appalachians is named after it. The river flows from Georgia into Tennessee, and the passion flower was recognized as the state wildflower of Tennessee in 1919.

- Pre-soaked seed may be sown in plug trays in spring, at about 68°F (20°C).
- Propagate by semi-ripe cuttings in summer.
- This perennial vine prefers sandy soils, in full sun to partial shade, and will tolerate drought.
- Passion flower requires frost-free conditions to produce good growth, flowers, and fruits.
- Mulch the roots to protect the plant in colder regions.
- Harvest the aerial parts for drying in mid to late summer.

The ripe fruits, known as "maypops," are made into jams and jellies. Commercially, they are used with those of *P. edulis* in juices and soft drinks. Juice freshly pressed from the fruits is used in cocktails and fruit salads.

Do not take while pregnant because it can cause the uterus to contract, or while taking antidepressant drugs.

Passion flower is calmative and sedative and clearly alleviates anxiety, nervous tension, stress-related disorders, and insomnia, as well as irritable bowel syndrome. It is also supportive for those undertaking withdrawal from addiction to narcotics.

If . . . life emits a fragrance like flowers and sweet-scented herbs . . . that is your success.

HENRY DAVID THOREAU, "HIGHER LAWS," *WALDEN* (1854).

Pelargonium spp.

Scented Geranium

Although commonly and wrongly called "geraniums," these popular plants belong to the *Pelargonium* genus and are mainly native plants of southern Africa. There are around 250 species, with numerous varieties of leaf scents, shapes, and sizes, and flowers of differing shades, from pink through to white. Scented pelargoniums were introduced to England and parts of Europe from Cape Province in the seventeenth century, but they did not become popular until the mid 1800s when their potential was recognized by French perfumiers.

Scented pelargoniums contain a large number of different complex volatile oils, and it is these that cause the plants to emit fruity or spicy fragrances, including orange, lemon, rose, mint, and balsam, and also blends of two or more of these scents. Among the most popular cultivars are *P.* 'Graveolens,' the rose pelargonium, and *P.* 'Attar of Roses,' both of which smell of rose water (or Turkish delight).

"Oil of geranium," derived from *P.* 'Graveolens,' has an uplifting and soothing aroma and is used extensively in aromatherapy as a treatment for various ills. The oil has also become an important base tone in perfumery, for both women's scents and men's colognes.

- As tender perennials, scented geraniums are usually grown in pots because they are not frost hardy.
- Like all the pelargonium family, they are easy to grow from cuttings taken in late summer and kept under glass with bottom warmth, good light levels, and a reasonably dry atmosphere.
- Numerous fine cultivars exist, such as 'Chocolate Peppermint,' 'Prince of Orange,' and even one named 'Cola Bottles.'

Fresh scented pelargonium leaves may be added to wine vinegar to make a distinctive base for vinaigrette dressing. They may be used to flavor jams, jellies, milk-based puddings, custards, ice cream, and dessert sauces, and to make herb teas, fruit punches, and summer spritzers. A layer of leaves laid at the base of a cake pan before baking adds flavor to a cake.

The wonderful and varied fragrances of scented pelargoniums may be experienced in different ways around the home. Leaves from mixed or single varieties may be used in potpourri, or infused in water to fill finger bowls at the table. Oil scented by the leaves may be added to beeswax furniture polish. The scented oil can also be used in room fresheners and burners.

Scented pelargoniums grown in the garden or in pots offer not only a variety of scents but also a wide range of leaf colors, shapes, and sizes. The popular *P.* 'Lady Plymouth' has gray-green leaves bordered in white.

Adding a few leaves, or a few drops of oil of geranium, to witch hazel (see p. 80) creates a scented skin toner for problematic skin. People with sensitive skin should opt for leaves because the oil can cause irritation.

Scented pelargonium oil is used in both aromatherapy and therapeutic massage as a treatment for depression; it is relaxing, soothing, and promotes a sense of well-being. Leaves are also infused as a vapor inhalant.

Perilla

Perilla is known as *zisu* in China, where it has been used medicinally since around 500 CE. The plant is thought to have originated in both China and mountainous areas of India; it was introduced into Japan in the eighth or ninth century CE. Its use as a vegetable and soup herb is mentioned in ancient Chinese recipes; while it is no longer favored in modern Chinese cuisine, it is still used in their medicine.

Perilla frutescens produces red and green leaves, known in Japan as red and green *shiso*; both are regarded as the quintessential Japanese herb for seasoning. *Shiso* is eaten with raw fish partly because it neutralizes potential stomach parasites in the fish. In Vietnamese cuisine, perilla is called *tai to* and used with rice vermicelli, and in spicy stews and dishes that are simmered for a long time. Aside from their culinary use, perilla leaves produced an oil used as oil-lamp fuel.

In England in the nineteenth century, perilla was mass-planted in public parks and gardens as a "bedding" foliage plant. Recently this practice has been taken up again, and the plant may again be seen in public parks and municipal places, often cleverly integrated into modern, nontraditional planting schemes. Perilla is thought to have been introduced into the United States in the mid 1800s, but it has now become invasive in some eastern and central states.

- Grow seed in modules under glass with a minimum temperature of 68°F (20°C).
- Transfer when root growth has developed, then plant outside or in pots after the last frost.
- A tender, bushy, annual, those with red-color foliage are the stronger growing, more commonly used form.
- Grow in rows for cropping as baby leaf, salad leaf, and mature leaves for cooking.
- Nip out the growing points to gain bushier plants.
- Red perilla loses color in cooling conditions.

In Japanese cuisine, perilla is used as a garnish with raw fish *sashimi*, and as a wrap and vegetable. It is served with bean curd, cucumber, and shredded cabbage. The red form is preferred because it turns juices pink. It accompanies *Myogi* ginger flowers and ginger thinly sliced as a pickle, and *Umeboshi*, the famous sour plum pickle. Leaves and flowering shoots are cooked in tempura batter with other vegetables, and they are added to soups. Seeds are salted and used as garnish.

Oil used from perilla leaves is added to tobacco and confectionery, and used in dental products. Oil pressed from the seeds is used in paper making, printing, and the paint industries.

Eating perilla seeds is said to protect intelligence and brain fitness, and to reduce cholesterol. The seeds are also used to introduce water to the digestive tract and thereby ease constipation.

Perilla is antispasmodic, antibacterial, and expectorant in its effects. It may be infused as a tea to treat colds, chills, bronchitis, and asthma, to alleviate nausea and abdominal pain, and to ease symptoms related to food poisoning and allergies, especially when caused by seafoods. However, perilla is grown and consumed in large quantities mainly because it is a powerful antioxidant, that is, it inhibits reactive oxygen and so slows aging.

Bistort

A native herb of parts of Europe, Asia Minor, and central Asia, bistort is naturalized in America. It is found growing in damp meadows, along natural water courses, and in old herb gardens. The common name derives from the Latin specific name *bistorta*, meaning "twice twisted," a reference to the growth habit of the rootstock. For the same reason it is known as "snake root" and "snakeweed"; indeed, bistort was traditionally used to treat snakebite.

Since at least the Middle Ages, the herb was cultivated as a valuable culinary and medicinal plant. In times of famine, the roots and leaves were eaten as emergency foods, particularly in Russia and Siberia. The roots contain a great deal of starch and were made into bread or roasted to feed the hungry.

Bistort roots are known to be among the most astringent in the plant world. They are very rich in tannin, which led to their use in treating the plague, smallpox, and fevers in the sixteenth and seventeenth centuries. The herbalist Nicholas Culpeper wrote that bistort could "help jaundice, expel the venom of the plague, smallpox, measles or other infectious disease, driving it out by sweating." It continues to be valued for its blood-staunching properties in the treatment of wounds.

To make Easter Ledges pudding, a revived savory recipe from the North of England, young bistort leaves are shredded and mixed with nettle leaves and other spring herbs, then baked on an oatmeal and barley base. Bistort leaves also go into salads, and the roots can be roasted after being thoroughly washed and soaked in water.

According to folklore, dwellings may be cleared of ghosts by burning the root as an incense, or by making an infusion and sprinkling it around the home. Doing so would also block the arrival of supernatural visitors.

Bistort is anti-inflammatory and used internally for diarrhea, colitis, dysentery, and excessive menstruation. The herb is used externally as a styptic for wounds, hemorrhoids, gum disease, and persistent bleeding.

- Can be grown from either divisions of the root stock or rhizome cuttings in the spring and fall.
- This hardy perennial likes moisture-retentive soil and prefers a sunny or semi-shaded site.
- The flower spikes make it an attractive garden plant.
- Cut back after flowering.
- Keep plants in check, because they can become invasive.

Vietnamese Coriander

Native to Southeast Asia, Vietnamese coriander grows profusely in the wild, mostly in the damp, moisture-retentive conditions of ditches, stream banks, and pond surrounds. Plants of the genus *Persicaria* are known as smartweeds or knotweeds and are related to the notoriously invasive Japanese knotweed, *Fallopia japonica*. But this knotweed is tender, not frost hardy, and is best grown as an herb in pots in temperate climes.

P. *odorata* is commonly known in Asia as "laksa leaves" or "Vietnamese mint," and it tends to be used like common mint. Laksa is also the name of a spicy noodle dish, favored by Malaysians and Chinese living in Singapore, which is topped off or garnished with chopped leaves of the herb.

In Vietnam, this coriander is traditionally cultivated and consumed by monks because it is believed to reduce sexual urges, thus making it easier for them to live a celibate life. (Conversely, the Vietnamese traditionally eat raw bean sprouts to enjoy the opposite effect.)

In Australia, research is taking place into the potential of an essential oil of the plant, kesom oil, for use by the food and fragrance industries.

Vietnamese coriander is added to steam baths, both to treat skin disorders and to improve and maintain the condition of healthy skin.

The herb is widely used in Vietnamese, Thai, Malay, Cambodian, and Australian cuisines for its hot, spicy, pungent, lemony taste. It goes into spring rolls, hot-and-sour soups, coconut milk-based dishes, and salads. It is often teamed with lemongrass for chicken, pork, and shrimp.

Antibacterial, anti-inflammatory, astringent, antioxidant, digestive, and tonic, Vietnamese coriander is used externally for the treatment of skin problems, such as acne, ringworm, and sores. Internally, it is used for indigestion, gas, and food poisoning, being effective against gut bacteria such as *Escherichia coli*.

Vietnamese coriander is thought to reduce fertility in women, so do not use it if planning to conceive.

- Propagate by taking cuttings in late spring.
- Cuttings will easily develop roots in water.
- The tender perennial plant prefers a fertile soil.
- It requires warm, damp, humid conditions to thrive. In temperate climates, a kitchen windowsill may be the best place for it.
- Cut back the plant if it becomes thin and straggly.
- Invigorate the plant and encourage new growth by feeding or re-potting.

Petroselinum crispum

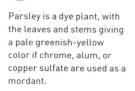

Parsley

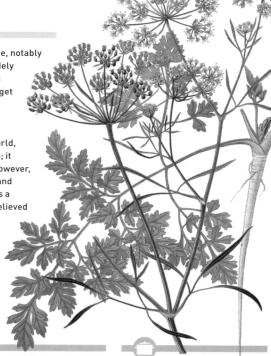

Parsley is native to central and southern Europe, notably the Mediterranean region, and is cultivated widely throughout the world. Breeders have produced numerous flat-leaf and curled forms, trying to get the best taste, vigor, and cropping excellence.

The herb was revered by the ancient Greeks, who held it as sacred to the dead. Parsley was dedicated to Persephone, queen of the underworld, and parsley garlands were used to honor tombs; it was also planted at burial sites. The Romans, however, were happy to eat it, using it liberally in salads and sauces. It was they who discovered its virtues as a breath deodorizer after garlic is eaten. It was believed to allay intoxication, and drinkers would wear sprigs on their heads to inhibit inebriation.

Many folk tales center on parsley, such as "It goes down to the devil seven times afore it grows"; "Parsley will grow where the wife wears the trousers"; and "Never transplant parsley, for it will always bring bad luck." Even death might be caused by disturbing it.

- Seeds are best sown into plug trays as parsley responds badly to root disturbance.
- Parsley is a biennial herb but it is best to sow seed annually to guarantee continual leaf production.
- For quick germination, bottom heat is needed for two to three weeks; otherwise it will be four to six weeks.
- Transfer outside when the soil is warm and moist.
- Parsley is a greedy plant; the soil should be rich in nutrients and not acidic.

Parsley is a dye plant, with the leaves and stems giving a pale greenish-yellow color if chrome, alum, or copper sulfate are used as a mordant.

An infusion of fresh parsley leaves may be used for anemia, poor digestion, stomach cramps, and gas. It will make a fine tonic as it contains many vitamins and minerals, including iron.

French and Italian parsley varieties have different tastes and should be selected for those cuisines. Add it to a cooked dish moments before it is served.

A hair rinse made from the seeds is said to kill head lice; a rinse from the leaves will darken hair and allay dandruff. A tonic and conditioner, parsley is also used to eliminate freckles.

Phlomis frusticosa

Jerusalem Sage

Jerusalem sage is a hardy perennial shrub native to the eastern Mediterranean region—Italy, Sardinia, Albania, Malta, Cyprus, and Greece into Turkey. It grows wild in rocky, wilderness areas, especially limestone hills, and has been cultivated in gardens for more than four hundred years; it remains a very popular ornamental plant. The herb was mentioned by Discorides in his *Materia Medica*, but otherwise little is known about its history.

The genus name *Phlomis* comes from the Greek *phlomos*, meaning "woolly plant"; *fruticosa* means "shrubby." *Phlomis fruticosa* belongs to the Labiatae, the same family as the common sage (*Salvia officinalis*, see p. 168), and is known as a sage because its leaves resemble those of common sage. The silver, velvety foliage has a reverse of almost white; the flowers are deep yellow.

Jerusalem sage is drought tolerant and is widely used in commercial landscaping. In Britain it is among the herbs that carry the prestigious Award of Garden Merit of the Royal Horticultural Society.

Although Jerusalem sage is not currently used by practicing medical herbalists, new research is being undertaken into the essential oil of the herb for its antibacterial and antifungal properties.

The leaves are strongly aromatic and dry well for potpourri. They were once used to make lamp wicks.

Jerusalem sage is not widely used as a culinary herb. However, in Greece the leaves are collected from the abundant wild plants that grow there, dried, and added to a mix of similar aromatic herbs for sale and use in local dishes. It is good in substantial stews and casseroles. Just one or two of the strongly fragrant leaves are enough to make a refreshing tea.

> *Yellow flowers bloom in a series of ball-like clusters known as verticillasters.*
>
> "PHLOMIS: JERUSALEM SAGE" ON DIGGINGDOG WEBSITE

- Propagate from softwood cuttings taken from non-flowering shoots in the summer.
- It is a good, dense shrub for providing ground cover, and it seems to be disease free.
- Cut back after flowering for more vigorous growth.
- Tolerates city pollution.
- An alternative species often planted is *Phlomis russeliana*, with flowers of a paler yellow and mid-green leaves.

For the dropsie,
fill an old cock with
Polipody and Aniseeds
and seethe him well,
and drink the broth.

WILLIAM LANGHAM,
THE GARDEN OF HEALTH (1683)

Anise, Aniseed

Anise is native to the Middle East, Egypt, Asia Minor, and Greece, and has been cultivated for thousands of years. It was a favorite of the Romans and was grown throughout the Roman empire, from the Mediterranean to England. At the time of Dioscorides, in the first century, the anise grown on Crete was considered to be the finest, its seeds rivaling any grown in Egypt.

In the Middle Ages, anise was called upon for its aphrodisiac properties as much for its use as a culinary spice or a calmative medicine; one source names it as one of the herbs that made up a "goodly love potion." It was one of the herbs that Charlemagne ordered to be grown in his gardens, and King Edward IV of England had sachets of anise and orris (iris) root (see p. 94) placed in his linen.

By the middle of the sixteenth century, anise was being cultivated in English gardens. It is still grown, especially in Europe and countries bordering the Mediterranean, as a flavoring for baked goods and confectionery. On a much larger scale it is also used in liquors, known variously as ouzo, anise, anisette, and pastis, and cordials, and as a flavoring for cosmetic and pharmaceutical products, including toothpaste. To supply these various industries, anise is cultivated in India, Spain, Italy, North Africa, and South America, although a different plant, star anise (*Illicium verum*), has replaced it in some cases.

- It takes a really hot summer for the seeds to fully ripen on the plant. Seeds will turn a grayish green color when they are ripe.
- In April, sow seeds in plugs, planting outside when the roots have established.
- Anise needs light, well-drained soil in full sun.
- If seeds are to be collected, wait until they are fully ripe, then cut down whole plants. Bunch the stems together and hang them head down over a bowl, or put the stems head first into a paper bag, so that none of the seeds goes to waste when they fall.

Anise tea improves the appetite after illness, and anise leaves and seeds can be infused as an aid to digestion, dispelling flatulence and calming and soothing the gut. Chewing the seeds after meals also helps to promote the secretion of digestive juices. Anise seeds are also reputed to be a galactogogue, that is, they increase the supply of milk in breast-feeding mothers, but excessive use of anise tea is said to impart an odor to the milk.

Chopped leaves are added to salads, and to cooked vegetables, soups, and stews before serving. They also enhance fruit salads, apple pies, gooseberry tarts, and pancake batter.

Anise is one of several herbs that were believed to have protective powers. Seeds were placed in bowls to ward off evil, and into pillows to keep away nightmares.

Mice and other small rodents keep well away from anise, so it is worth putting it down in cupboards, storage sheds, and seed glasshouses, as a repellent. Insects, too, give anise a wide berth.

The Romans baked anise seeds into special cakes that were served at the end of a wedding feast. These were an early version of the traditional wedding cake.

Pepper

The pepper is a rapidly growing woody-stemmed vine, native to the equatorial forests of Asia and India. Today it is additionally cultivated in tropical countries such as Brazil, Indonesia, and the West Indies. The vine produces long racemes of small white flowers in clusters; these give way to the fruits that are harvested and dried as peppercorns.

Pepper is one of the oldest known spices and was being used in India 4,000 years ago. Hippocrates prescribed it as a medicine, and it was used as currency in payment of levies and taxes. In 408 CE Alaric, king of the Visigoths, demanded 3,000 pounds (1,360 kg) of pepper as part of a ransom for the City of Rome. It was the most important commodity traded through the Middle Ages between India and Europe, making Venetian and Genoese merchants extremely wealthy. In England "peppercorn rents" were paid in weight of pepper, and are still referred to today. The pepper trade, dominated by the Portuguese until the eighteenth century, was fought over by the European sea powers.

Three colors of *Piper nigrum* peppercorns, green, black, and white, are sold today. All occur on the same plant, being the same fruits but harvested or processed at different stages. Green are the soft, unripe fruit, often preserved in brine but also dried. Black are the sun-dried red berries, picked before they are ripe and dried until black. White peppercorns are the hard seeds extracted from fully ripe red berries.

- Pepper is a tropical plant and a mature specimen cannot be grown in a temperate climate except inside a heated glasshouse.
- Sow seed at 63–77°F (20–25°C) in early spring, or take semi-ripe cuttings in summer.
- The plant requires a rich soil and high humidity.
- Vines do not produce a crop until they are seven or eight years old. A mature vine may be 20 feet (6 m) long.
- The vine is susceptible to pests including the pepper flea beetle, the pepper weevil, and fungal root rot.

With salt, pepper is the condiment most used in the Western world—most dining tables are set with the duo. It is included in many savory dishes, meat products (especially sausages), sauces with cream for steak, salad dressings, pickles, and as a coating for meats, fish (especially smoked), and cheeses. Green peppercorns go into pâté and sauces and work with roast duck, stock, and soups.

Black pepper has been successful in inhalant form as a smoking cessation treatment, partly because it gives the sensation of smoke in the respiratory tract.

Black, white, and green peppercorns are sometimes mixed in pepper grinders for decorative effect. Pink peppercorns, from the tree *Schinus terebinthifolius*, may be added for extra color.

Pepper is used in Western, Chinese, and Ayurvedic medicines as an antibacterial, aromatic, stimulant, carminative, and digestive agent. It is used to stimulate gastric juices, aid digestion, and dispel gas. The Chinese use it for treating food poisoning, especially from fish, shellfish, and meat. In aromatherapy it is used diluted to promote blood circulation, ease sore muscles, and flush away toxins. It may be used with a diffuser to relieve colds and flu.

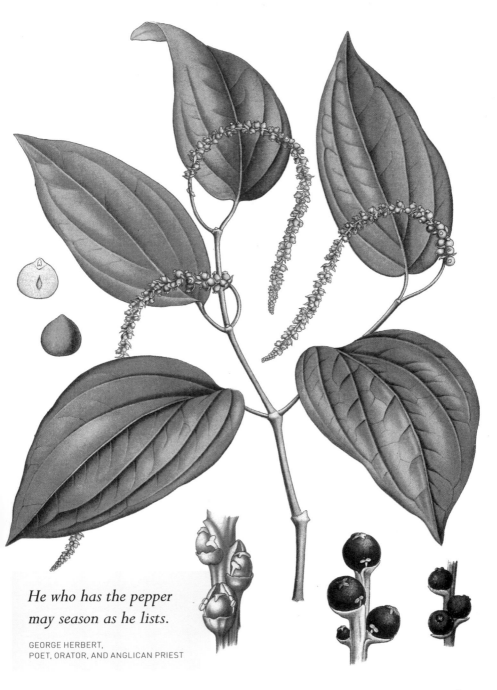

He who has the pepper
may season as he lists.

GEORGE HERBERT,
POET, ORATOR, AND ANGLICAN PRIEST

Plectranthus amboinicus syn.*Coleus amboinicus*

Mexican Mint

Plectranthus amboinicus is native to southern and eastern Africa, South Africa, Swaziland, Angola, Mozambique, Kenya, and Tanzania. It is cultivated for its essential oils in many tropical regions, including the Caribbean, Southeast Asia, Java, Cambodia, Vietnam, southern India, and Mexico, where it is also used as an ornamental and bedding plant. Its widespread culinary use is reflected by its geographically and botanically diverse common names, which include Mexican mint, Indian borage, Cuban or French oregano, and Spanish thyme.

Mexican mint has been used traditionally as a medicine by many cultures, and it is widely grown in domestic gardens for home treatment of malarial fevers, bronchitis, colic, and epilepsy; it is also effective for wounds and scorpion bites. In Cambodia, juice pressed from the leaves is given to children to protect them from the common cold.

The people of Brazil, and especially immigrants from Africa, would burn the herb to consecrate their temples, as well as in magical rituals enacted to cleanse the home and invite protection from harm. The herb was also thought to increase psychic abilities.

- A tropical perennial plant, Mexican mint is propagated by cuttings taken in late spring or early summer.
- The plant is not frost hardy—it is often grown in pots and brought indoors in the cold season.
- Planted outside, it prefers a shady, warm position with good light levels. Full sun will scorch the leaves.
- Does not like to dry out, but is easily over-watered.

Adding a few drops of the essential oil of Mexican mint to the final laundry cycle can give clothes and linen a fresh, minty smell. Used on the hair, it treats dandruff.

Mexican mint is anti-inflammatory and antiseptic, and so used to treat skin eruptions, wounds, bites, sores, and burns. It is thought to be beneficial for the liver, and is a remedy for coughs, colds, and headaches.

Strongly aromatic, the herb is used for Jamaican jerk chicken and salt cod, and bean dishes, especially Cuban black-bean soup. It is added to tomato and tomatillo salads, vegetable dishes, and salsa, and flavors beef, lamb, and game. It is often used as an oregano substitute.

[Its Chinese name means] "giving fragrance to the hands."

GREENCULTURE WEBSITE

Jacob's Ladder

Also known as Greek valerian, Jacob's ladder is a herb native to northern and central Europe, Siberia, and North Africa; it was introduced to North America, where other members of the genus can be found. There is reference to it growing wild in certain parts of England, especially liking moisture-retentive habitats, but it is now rare there.

The generic name *Polemonium* is somewhat puzzling: it derives from the Greek *polemus*, meaning "war," but there is no known explanation as to why. The common name, Jacob's ladder, refers to the ladderlike arrangement of the leaves.

The herb was a favorite with ancient Greek physicians, who recommended it to be taken with wine to relieve toothache, dysentery, and insect bites. *Polemonium* species in general were used by Native American tribes as a hair wash, while the Thompson and Meskwaki tribes saw it as a "powerful physic" to treat hemorrhoids. It was once called *Herba Valerianae Graece* and used to treat syphilis, rabies, and neuralgia. The herbalist Nicholas Culpeper thought it useful for malignant fevers and pestilential distemper.

The plant is also called "cat valerian" because cats seek it out, like they do catmint (*Nepeta cataria*, see p. 123).

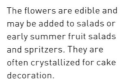

- Sow seeds in plug trays in the fall or spring, and plant outside when the roots have formed.
- Jacob's ladder is a hardy herbaceous perennial that likes rich, moisture-retentive soil, and a sunny or semi-shaded position.
- The pretty, cup-shaped, lavender blue or white flowers appear in the early summer.
- The plant is likely to spread by self-seeding.
- Bees are highly attracted to the flowers.

The flowers are edible and may be added to salads or early summer fruit salads and spritzers. They are often crystallized for cake decoration.

In the past, Greek valerian was used for a wide range of conditions, from fevers, heart vapors, and hysteria to epilepsy, but the herb lost favor and is rarely used medicinally today.

And [Jacob] dreamed, and behold a ladder Set up on the earth, and the top of it reached to heaven.

THE BIBLE, GENESIS 28:12

Polygonatum odoratum syn. *P. officinale*

Solomon's Seal

This herb's genus name, *Polygonatum*, comes from the Greek for "much-jointed" and, as the doctrine of signatures directed, this characteristic led to its being used to treat joint disorders. Herbalist Nicholas Culpeper commented that "Saturn owns the plant, for he loves his bones well," and recommended it "to knit any joint, which by weakness uses often to be out of place, or will not stay in long when it is set." John Gerard would also recommend it for use for similar ailments, "as touching the knitting of bones," writing that "there is not another herb to be found comparable" for sealing (healing) wounds.

Solomon's seal, also called St. Mary's seal, is native to woodlands of Europe. There are various schools of thought regarding the origin of those common names. One explanation is that the flat, round scars on the outside of the rhizome resemble impressions of a seal; another is that the rhizome, when cut transversely, reveals figures resembling Hebrew characters. Again, Solomon was a king who understood medicinal herbs and knew their value, and it could be that he was known to have set his seal of approval on this particular herb.

In Turkey the young shoots were once boiled and eaten, in the way of asparagus. The roots, meanwhile, are rich in starch and have been used as a subsistence food in times of famine. The dried flowers and roots were once used as a snuff to relieve headaches.

All parts of Solomon's seal, especially the berries, are potentially harmful. Even in small quantities for medicinal purposes, the plant should only be used by a qualified practitioner.

In Ayurvedic medicine, the flowers and roots were used to make a rejuvenating tonic with aphrodisiac properties. The herb has long been employed as an ingredient of love philters and potions.

- Grow from seed sown in the fall, in pots placed outside.
- Germination can take twelve to eighteen months.
- Plant in the soil when the roots are established.
- May also be propagated by root division in early fall and early spring.
- This hardy herbaceous perennial prefers light, fertile soils and a shady site.
- Cut back after flowering, and feed.

In China the plant is called jade bamboo, due to its bamboolike leaves. It is used in Chinese medicine for tuberculosis and heart disease, and is associated with the heart, kidney, lungs, and spleen. In Western medicine it was used for coughs, bruises, and hemorrhoids. In Ayurvedic medicine it is used for infertility, and added to warm milk and ghee (clarified butter) as a tonic.

Portulaca oleracea

Purslane

Purslane is native to wide swathes of Europe and Asia and has been introduced to many countries around the world for cultivation as a herb crop. Its succulent, fleshy leaves are highly nutritious and contain a wealth of vitamins and minerals. It is a measure of purslane's value that chickens feeding on it produce eggs rich in omega-3 fatty acids.

In some ancient cultures, purslane was considered an "anti-magic" or magic-neutralizing herb, and so it was strewn around the bed to ward off evil spirits and prevent nightmares. Even in comparatively recent times, soldiers carried it into battle as a good luck charm that would protect them from harm. In China, medical records dated to around 500 CE describe how the plant was used medicinally. For many centuries it has been cultivated as a potted herb in India and the Middle East, where it is often consumed as a component of salad.

In North America purslane was known to Native American tribes, with the Cherokee using it in a decoction for intestinal worms. The Iroquois cooked it as a vegetable, but also saw it as an antidote to "bad medicine"; they would also use pulped leaves to cool and heal burns. The Navajo, in turn, used purslane as an analgesic and potted herb.

• Purslane is a half-hardy annual that is usually grown from seed sown in the spring.

• It is quick to germinate, and if conditions are right it can be sown successively at monthly intervals for cropping as a herb or vegetable for use in cooked dishes or salads.

• Purslane prefers light, rich soil with ample water and full sun.

• Both the stems and leaves can be harvested and eaten.

Young, sweet leaves are used in fresh leaf salads; older leaves in clear vegetable broths and soups. The whole tops of the plant are bundled, stood upright in a jar, and picked in vinegar. With lemon and sumac, purslane is an important ingredient of the Middle Eastern dish *fattoush*, a sour but delicious salad.

Purslane juice is taken internally to treat dysentery, mixed with honey for coughs, and was once recommended for breast inflammation. Externally, the bruised herb was applied to the forehead and temples to "cool and temper." Culpeper used it for thirst in illness, to ease the heat and pain of gout, and also for inflamed eyes.

Anyone collecting purslane in the wild should beware of spurge, a poisonous plant that often grows alongside it. Purslane stems are thick and contain clear water; spurge stems are thin and wiry and contain a milky, poisonous sap.

*Tisty Tosty, tell
me true, who shall
I be married to.*

FOLK RHYME

Cowslip

The cowslip is a much-loved herb native to northern and central Europe. It grows wild in meadows, old pastures, fields, and on the edges of woodlands. In addition, it is often cultivated in semi-natural wildlife meadows, cottage borders, and by small, man-made woodlands. It was one of the flowers grown in Elizabethan knot gardens.

The common name of cowslip comes from the Old English *cuslyppe*, meaning "cowpat," or cow dung. The generic name *Primula* comes from the Latin *primus*, meaning "first," a reference to the early spring appearance of the flowers. In the English countryside, young people would make a ball out of cowslip flowers, called a "tisty tosty," for a game in which they would learn of future loves. Children had another use for the flowers: they would suck the delicious nectar from their necks as they walked to and from the village school.

Another common name for the plant is St. Peter's keys; this refers to the pendent flowers, which dangle like the keys held by St. Peter as he waits at the gates of heaven. The plant was said to have been a favorite of nightingales, which were thought only to frequent places where they grew. Fairies, too, were said to hide where cowslips grew.

The plant known to some Americans as the cowslip is not *P. veris* but *Caltha palustris*; in Britain this is known as the "marsh marigold."

- Likes a wild habitat best.
- Seed should be sown fresh to guarantee good germination.
- Collect seed heads and sow directly onto the compost surface in trays or pots, cover with glass, and place in a cold frame.
- After four to six weeks, remove the glass. Seeds need first cold then warm temperatures to germinate.
- This hardy perennial will gradually self seed if the flower heads are left on.
- Grow your own plants; do not be tempted to take them from the wild.

Young leaves and flowers are eaten in salads. Flowers are used to color sweet desserts, and as a decoration on cakes. They may be preserved in sugar and eaten with fresh cream.

Cowslip is antispasmodic and anti-inflammatory. Tea from the flowers is good for insomnia, headaches, and nervous tension. The roots are used to treat bronchitis, dry coughs, and whooping cough. Salicylates in the roots are good for arthritis, a fact recognized in old herbals by references to the root as "radix arthritica." The herb is also known as "herba paralysis" and "palsywort" for its use in treating spasms and cramps.

Cowslips were associated with youth; a young person would carry one to preserve their youth, while older people would do so to bring youth back. Cowslip was also associated with healing.

Cowslip wine is a delicately flavored, amber-colored drink made from the nectar of the flower. Once common, it is now made less often because plant populations have been reduced.

Although Nicholas Culpeper recommends cowslip for enhancing beauty and for treating spots, wrinkles, sunburn, and freckles, it can provoke contact dermatitis in susceptible individuals.

Primrose

In the language of flowers, primrose means "believe me." It is a symbol of early youth, innocence, and the doubts and fears of young lovers. Primrose is the birthday flower of people born on the 7th of May.

This native flower of Europe, Asia Minor, and North America is found growing wild in neglected meadows, grassy banks, hedgerows, and woodlands. The generic name derives from the Latin *primus*, meaning "first," and so the name is a corruption of *prima rosa*, or "first rose"; in Old French the name was *primerose*.

In 1653 Nicholas Culpeper wrote: "The juice of the roots stuffed up the nose occasions violent sneezing, and brings away a great deal of water, but without being productive of any bad effects." Herbalist John Gerard recommends that primrose tea be drunk in the month of May as "famous for curing phrensie [frenzy]."

In the seventeenth and eighteenth centuries, when the primrose was more abundant, the flowers were collected for many culinary treats; for example, they were candied, pickled, added to vinegar, and made into wine.

- Fresh seed, still green and not brown and dry, may be sown in summer. Germination takes a few weeks to occur.
- Dried seeds needs stratification. Sown in the fall, they can take up to two years to germinate, so patience is needed.
- Division of mature plants in the fall may be more successful than sowing seed.
- Primrose is a hardy perennial and will spread.

Primrose is antibacterial and anti-inflammatory, and is used for respiratory infections such as bronchitis, and to cleanse wounds. It is a sedative and is used for insomnia or anxiety.

Flowers and young leaves may be eaten in salads. The flowers are used for decoration on cakes and buns, and as a flavoring for desserts. Primrose pottage was made by boiling pounded flowers, honey, almond milk, saffron, rice flour, and powdered ginger.

Aske me why I send to you This Primrose, thus bepearl'd with dew?

ROBERT HERRICK, "THE PRIMROSE"

Plants of the *Primula* genus can cause allergic contact dermatitis in some individuals. *P. vulgaris* is relatively safe, but it is best to handle it with care.

Prostanthera spp.

Australian Mint Bush

Prostanthera is a genus of highly aromatic, usually bushy, evergreen shrubs native to Australia. The ninety species of this Australian herb are members of the Lamiaceae/Labiatae, or mint family. They grow in wild places of New South Wales, Victoria, Tasmania, and South Australia, on sheltered rocky hillsides and the edges of rain forest–dominated areas.

    Prostanthera cuneata, the alpine mint bush, is one species that is often available in temperate countries outside Australia. It has small, wedge-shaped leaves containing oil glands; when the leaves are crushed, the glands emit a volatile oil with a scent recalling both mint and eucalyptus. *P. cuneata* has beautiful, small, orchidlike white flowers with purple, yellow, and red blotches within the throat. It is very attractive to bees. *P. rotundifolia*, roundleaf mint bush, is probably the hardiest of the family, but may still require some protection. Growing to 6 feet (2 m), it has attractive, bell-shaped, purple flowers and round to ovate aromatic leaves. This species is used by the Native Aborigines as a bush food, or "bush tucker"; other species are mainly used as a means to light fires.

- The genus can be grown from seed but is slow to germinate in cooler climates.
- Propagation is better from softwood cuttings taken in early summer.
- The plants prefer a loam-based compost and a well-drained, sunny position.
- The perennial evergreen shrubs are tender outside southern Australia. They need frost protection throughout the winter.
- They make very attractive pot or container plants.
- Prune lightly after flowering.

Prostanthera species were used historically to spice up a bush food called "bunya nut butter." Today, bush-food specialists use them in cordials, syrups, chutneys, and salsas, and also use them to preserve foods.

The essential oils of *Prostanthera* species are unexpectedly powerful. Exercise caution if using them for medicinal or culinary purposes.

Prostanthera is antibacterial, antifungal, and carminative. Once used by the indigenous people of Australia, it is not found in the Western pharmacopoeia. *P. rotundifolia* may be taken internally for headaches and colds, and used externally as an inhalant. *P. cuneata* is sometimes used as a flower essence for purposes of emotional care.

Self-heal

A wild herb found growing in abundance throughout the Northern Hemisphere, self-heal is native to Europe, Asia, and North America, and was introduced to China and Australia. It prefers well-drained hedgerows, woodland clearings, grasslands, and pastures, and in the human-dominated environment it can spread rapidly in lawns. By those who do not appreciate its role as a wildlife habitat it may be regarded as an invasive nuisance, although it is easily controlled. The steady march of the plant is facilitated by its creeping fibrous roots.

Self-heal's flower resembles a throat, and so, according to the sixteenth-century Doctrine of Signatures, it was previously used as a treatment for numerous throat complaints. Indeed, the genus name *Prunella* has nothing in common with *Prunus*, as might be imagined, but is instead derived from *Brunella* or *Braunella*, the German for "quinsy," a serious inflammation of the tonsils resulting in abscesses.

Nicholas Culpeper explained the name of self-heal as "when you are hurt, you may heal yourself . . . it is an especial herb for inward and outward wounds." This traditional use also explains another of its common names, carpenter's herb, because it was used for bruised or cut fingers.

- This hardy perennial can be grown from seed sown in spring or fall.
- Division is the easiest form of propagation. Runners have small root systems and may be cut and planted in spring or fall.
- Self-heal can be persistent as a weed in lawns, and may grow very prolifically once established.
- A good wild-habitat and nectar plant that flowers from mid to late summer and is attractive to bees.

Young leaves may be added to a leaf-and-wild-flower salad, and also used as a flavoring for soups and stews. The leaves may be infused as a tea that is served either hot or cold.

Self-heal mouthwash is used for ulcers, oral eruptions, and cold sores (*Herpes simplex*). In World War I, self-heal was chewed in the trenches to prevent gum disease.

Self-heal is antibacterial, antibiotic, anti-inflammatory, antioxidant, astringent, diuretic, and detoxifying in its effects. In Chinese medicine it is used as a liver and lymphatic tonic, and as a treatment for various abdominal swellings. In the West it was regarded more as a wound herb, for application to cuts, abrasions, and sprains. It remains an effective skin wash for bruises, burns, and sores.

Pulmonaria officinalis

Lungwort

This native herb of mainland Europe, Denmark, Sweden, and North America is also found naturalized in woodlands and thickets of the British Isles, central Russia, and northern Italy.

Lungwort is cited as the best example of the Doctrine of Signatures, which held that God suggested the medical uses of plants by including visual clues in their appearances. In its maturity, lungwort has mottled leaves that resemble diseased lungs, and accordingly the herb was used to treat coughs, bronchitis, and other respiratory conditions. The original concept was developed by Paracelsus (1493–1541), who wrote that "Nature marks each growth . . . according to its curative benefit."

Numerous other folk names refer to physical attributes of *Pulmonaria officinalis*. For example, the name "Virgin Mary's tears" interprets the white spots on the leaves as tears; also, the flowers change from pink to blue, with sprigs bearing both pink and blue blooms, and this was seen as Mary's blue eyes becoming red from weeping during the crucifixion of Christ. Another common name, "soldiers and sailors," also refers to the two colors, although this time they were likened to the red and blue uniforms of the British army and navy.

Picked early in the spring, young leaves may be added to linden leaves and arugula to make a healthy tonic salad rich in vitamins B and C.

P. officinalis flowers early and its nectar is attractive to bees. In Russia the plant is known as *medunitza*, while the Polish call it *miodunka plamista*; both terms mean "honey" and refer to the plant's importance to bees and honey production.

Anti-inflammatory, expectorant, and mucilaginous, lungwort can be infused to treat sore throat, coughs, catarrh, bronchitis, and other respiratory conditions; its silica benefits the lungs. A lungwort eye bath restores sore, gritty, tired eyes.

Lungwort is a member of the Boraginaceae family, and the hairs on the leaves can cause an allergic reaction. Lungwort should not be eaten during pregnancy.

- Lungwort self-seeds freely and spreads quickly.
- Propagate by division after flowering in late spring or the fall.
- The hardy, perennial, semi-evergreen plant prefers well-drained soil that is moisture retentive.
- Cutting back leaves and stems after flowering improves the quality and vigor of the plants.

Pasque Flower

The pasque flower is a native of central and northern European countries, being found in chalk and limestone areas of England, France, Germany, Denmark, and southern Sweden. The plant flowers around Easter, hence the name "pasque," which means "like Paschal," a reference to Easter; the blue sepals were once used to dye Easter eggs.

The pasque flower appears in many legends. It was said that the flowers only open when the wind blows, and that the plant sprang from the blood of Adonis. In some parts of Britain the plant is known as "Danes' blood" because it is associated with places where their blood was shed, often in ancient barrows and sites where invading Danes were repelled by King Alfred.

A member of the buttercup family, the pasque flower has long been used medicinally; Galen and Dioscorides both praised its properties. However, knowledge of its use in ancient times has been obscured by confusion over its nomenclature. Linnaeus categorized it as *Anenome pulsatilla* and only recently was it renamed *Pulsatilla vulgaris*.

Juice extracted from the petals has been used as a green stain for linen and paper, although the color is not permanent.

Only qualified herbalists should use this plant. Analgesic, antibacterial, antispasmodic, and respiratory in its effects, it is used to treat asthma, bronchial infections, and whooping cough. It relieves certain types of headache and neuralgia, and also toothache and earache.

The fresh plant is highly toxic and should never be taken internally. It also causes contact dermatitis and skin irritation; gloves should be worn when handling it. Drying the plant changes its constituents, making it safer for medicinal use.

The drug Pulsatilla is obtained not only from the whole herb of *P. vulgaris*, but also from *P. pratensis*, the meadow anemone, which has smaller flowers and sepals of a deeper purple.

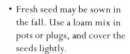

• Fresh seed may be sown in the fall. Use a loam mix in pots or plugs, and cover the seeds lightly.
• Placed in a cold frame, seeds of this hardy perennial can take as little as a month or as much as a year to germinate.
• The plant is attractive as an early-spring-flowering herb.
• Flowers will only appear in the second season of the plant's life.

Virginia Mountain Mint

The genus *Pycnanthemum* contains more than twenty species of highly aromatic plants, all known as mountain mints. The genus is native to North America and belongs to the family Lamiaceae/Labiatae; the plants are thus related to the true mints of the genus *Mentha*. Although *P. virginianum* is known as mountain mint, it actually grows in the moist soils of wet meadows and swamps, and alongside streams and ponds. As the name suggests, it grows in Appalachia, but it also grows wild prolifically in areas farther west, particularly Missouri.

The genus name comes from the Greek *pykos*, meaning "dense," and *anthos*, or "flower," a reference to the growth habit of the flower heads.

Native American tribes have long used the herb as both a food and medicine. The Chippewa used the buds and flowers to season meats and broths, and used a decoction of the leaves as a medicine for chills and fevers. The Lakota infused the plant for a cough treatment. The Meswaki and other tribes used it as a tonic and as an ingredient of a stimulant compound that was pressed into the noses of individuals near death in the hope of reinvigorating them.

- Sow seed in modules in spring, and plant outside when roots have developed.
- Propagate by division in early spring.
- The hardy herbaceous perennial prefers sun or semi shade in a moisture-retentive soil.
- The small, white to lilac flower clusters appear from July to September.
- This mint makes an attractive addition to a herbaceous border or herb garden, or to a wild garden or stream side.

Mountain mint is a tonic and is used to help patients recover after illness. It is used to treat colic, chills, and fever, to increase perspiration and improve digestion.

The minty leaves, flower tops, and buds enhance the flavors of dishes, soups, stews, salads, and meats.

Mountain mint is highly attractive to butterflies, bees, and other beneficial pollinating insects.

> *Mountain mint is the only common name for at least fourteen species.*
>
> DONALD D. COX,
> *A NATURALIST'S GUIDE TO FOREST PLANTS: AN ECOLOGY FOR EASTERN NORTH AMERICA* (2003)

The rose looks fair, but
fairer we it deem
For that sweet odor
which doth in it live.

WILLIAM SHAKESPEARE, SONNET 54

Apothecary Rose

The rose classified as *Rosa gallica* var. *officinalis* is thought to have been the original rose of antiquity. It has long been a symbol and token of love, and bunches found in Tutankhamun's tomb are thought to have been placed there by his young wife. The rose originated in northern Persia (Iran) and was brought across the Middle East to Egypt and Greece, and later to Italy and France. The Roman poet Horace (first century BCE) wrote at length on how to grow roses in specially prepared beds to ensure blooms of the best quality.

The name *Rosa* comes from the Greek *rodon*, meaning "red," and *R. gallica* is a deep crimson. In Classical myth, Venus loved Adonis, and she caused the flower to spring from his blood as he lay dying. The Romans lavished *R. gallica* blossoms and petals on any social event. At banquets especially the petals were added to food and drink and strewn on the tables and floors as decoration. Crowns of roses adorned the heads of brides and grooms, and petals were strewn before the feet and chariot wheels of victorious armies. The rose symbolized pleasure and indulgence to the Romans, but it also had a role at funerals. In addition, it was an emblem of secrecy and silence. For centuries it was carved into ceilings of rooms to remind guests that information was imparted *sub rosa*, or "under the rose," to be kept in strict confidence.

- This rose can be grown from seed sown in the fall, or by budding in the summer.
- Alternatively, it may be propagated from hardwood cuttings in the fall.
- This hardy, perennial, deciduous shrub prefers well-drained but moist, fertile soil, and a position in full sun.
- Most species and old roses flower on the previous year's growth, and therefore *R. gallica* var. *officinalis* should not be cut back hard.
- Feed the rose well after cutting back.

R. gallica var. *officinalis* is astringent, antibacterial, and antiviral in its effects. A compress of the buds and petals may be used to soothe and relax tension headaches, and ease sore, itchy, irritated eyes. A rose gargle or infusion can benefit sore throats, colds, bronchial infections, and stomachic and digestive ailments. Rose is restorative and uplifting on the nerves, and psychologically calming and soothing. Rose infusions can be helpful in alleviating depression and grief.

R. gallica var. *officinalis* buds are a component of particular potpourri formulations suggested for use in the bedroom and other romantic settings.

The scent of roses is one of the most popular additions to commercial preparations such as skin creams, bath oils, and salves. The popularity of aromatherapy has increased the use of rose oil in such products.

Like *R. damascena*, *R. gallica* is used to make rose water, a flavoring for Turkish delight, sorbets, ice creams, jellies, cordials, and a variety of desserts. The North African spice mix *ras el hanout* contains petal and buds. Petals are added to salads, fruits, wine, and summer punch.

Do not take during pregnancy because its safety is not clearly established.

Rosmarinus officinalis

Rosemary

Rosemary is native to the Mediterranean and was introduced throughout Europe before being taken farther afield. It is a one-species genus, but there are many varieties with differing leaves, flowers, colors, and growth patterns. The varieties differ also in aroma, with some having hints of lemon or ginger. The genus name is derived from the Latin *ros marnus*, meaning "dew of the sea," a reference to its natural seaside habitat.

The warm aroma of rosemary has been appreciated since ancient times. A symbol of fidelity, friendship, and remembrance, it has long been used for culinary, cosmetic, and medicinal purposes. In particular, throughout history it has been used to improve memory. Students in ancient Greece wore rosemary in their hair and garlands around their necks to help them through their scholarly exams. Nicholas Culpeper wrote that, "The sun claims privilege in it and it is under the celestial Ram (Aries). It helps a weak memory and quickens the senses." To this day, people inhale the vapors of rosemary oil as a memory aid.

At weddings, a bridesmaid would plant a sprig from her bouquet to be used by any daughter from the marriage on her wedding day.

- Seed needs bottom heat of 80–90°F (27–32°C) to germinate. Sow in plug trays in spring. Do not over-water.
- Propagation is much easier by cuttings taken in spring from the new growth.
- This evergreen, hardy, perennial shrub requires a well-drained sunny position.
- Rosemary is a good container plant. It is best to cut back after flowering.
- Sprigs may be gathered whenever required for cooking throughout the year.

Leaves are added fresh to roast meats, goose, rabbit, duck, potatoes, and barbecues, and used to flavor oils, marinades, and sauces. Chopped leaves go into herb butter and are baked into bread and scones. Flowers are sprinkled on salads, fresh fruit, and creamy cheeses, and on dishes as a garnish.

Rosemary may be infused as a tea, to be taken no more than three times a day and only for short periods. It improves mental clarity, energizes, improves circulation, and protects against free radicals. It is a liver stimulant and speeds up the removal of toxins. A bath in rosemary leaves or oil alleviates aches, pains, and fatigue, and a rub with rosemary oil will ease rheumatic and arthritic pains.

Rosemary was burned with juniper berries in sick rooms to allay the spread of infection and to purify the air. In the garden it makes a beautiful hedging plant, especially around a vegetable garden.

Rosemary is an antiseptic body moisturizer and cleanser, and is a good hair rinse, shampoo, and conditioner for brunettes. A facial steam will unblock pores. Rosemary mouthwash will freshen the breath.

Rosemary can be toxic and abortive in large doses, and so should not to be consumed in pregnancy or while breast-feeding. It is not recommended for anyone with high blood pressure, epilepsy, or seizures.

There's rosemary, that's for remembrance. Pray you, love, remember.

WILLIAM SHAKESPEARE,
OPHELIA IN *HAMLET*, ACT IV SCENE 5

Rubia tinctorum

Madder

Rubia tinctorum is native to southeast Europe and Turkey, while other plants of the genus *Rubia* grow in North and South America and large areas of Asia. A sprawling, weedlike plant, madder has a tangle of thick woody roots that are bright red inside. The roots contain alizarin and have been used for centuries to make a dye, sometimes known as "Turkey red." The dye has long been used for carpets, calicos and cottons, shawls and drapes, and many other natural-fiber textiles, and the madder plant was grown around Europe wherever cloth industries were formed: in France, Germany, Holland, and, in a small way, England. In the Middle Ages, madder and woad (which produces a blue dye, see p. 95) were both of great importance. Whole economies in Western Europe revolved around their culture and use until the plants were respectively usurped by imported cochineal and indigo.

Over the centuries, madder dye was produced by elaborate, almost ritualistic means. Numerous components were required to achieve the end result, and dyers were careful to keep their recipes secret. Ancient Hebrew laws only allowed the growing of madder for domestic use, and insisted that only wooden tools could be used to harvest it. Pliny the Elder, writing of madder growing near Rome in the first century CE, recommended it for treating jaundice.

- In spring or fall, sow seed, plant cuttings, or divide the roots of plants.
- Prefers well-drained, alkaline soil in full sun.
- Requires space to grow to its full potential and produce a crop of usable roots.
- Harvest after the third or fourth year, when roots will be mature and at their best.

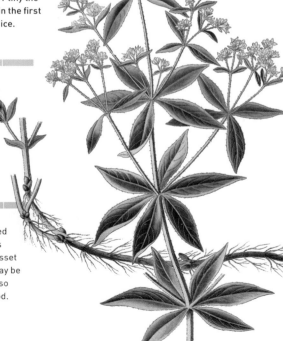

Madder, used powdered or as an infusion, is antiseptic, antibacterial, anti-inflammatory, diuretic, and laxative. It was once used to treat kidney stones, bladder complaints, and jaundice. The dye in the roots reappears in urine, sweat, and milk, giving them a pink hue.

Madder's main use is as a fabric dye, derived from the roots. Depending on the mordants used, it can produce pinks, corals, reds, russet reds, and oranges. Once dried, the roots may be stored in glass or stone jars; paper bags also suffice. The roots keep well for a long period.

Common Sorrel

Sorrel is native to Britain, Europe, and Asia, and is naturalized in the United States. In the wild it grows mainly in old grasslands and pastures, and on rich, damp, loamy, acid soils that contain iron. The common name is derived from *sur*, meaning "sour," and the Old French for the plant is *surele*. The sourness of the leaves and stems is due to the presence of oxalic acid. The levels of oxalic acid are high enough to make sorrel an ideal substitute for rennet as a means of curdling milk; sorrel juice is particularly used in Lapland for that purpose.

The ancient Egyptians and Romans used sorrel as a bitter counterpart to richness in their food. It was a popular salad and "pot herb" in the medieval period until it was supplanted by French sorrel, *Rumex scutatus*, which has larger, more luscious leaves. Apothecaries knew common sorrel as "Herba Acetosa" and included it in their remedies from the fifteenth to the nineteenth century. Sorrel has a high vitamin C content and was used to prevent or treat scurvy. The leaves are sometimes chewed by country people as a means of allaying thirst.

- Sow seed of this hardy herbaceous perennial under glass in early spring, and in rows outdoors in late spring.
- Seed is quick to germinate, within ten to fifteen days.
- Divide plants every other year in the fall. Doing so prevents development of a woody core and keeps the leaves succulent.
- Cut back flowers and feed to encourage leaf production.

A strong infusion of sorrel will remove from linen stains caused by rust, mold, or ink. Sorrel is also good for cleaning wicker or tarnished silver.

A green/yellow dye is obtained from the flowering tops and leaves.

Sorrel may be used as an alternative to spinach in potato, egg, leaf salads, soups, sauces, omelets, vegetables, casseroles, roulades, and cream cheese. Constance Spry recommends a sorrel wine. Sorrel loses its vibrant color when cooked. Cook with stainless steel—the leaves react badly to iron.

Sorrel is antiseptic and laxative but is seldom used today. It was applied externally for ringworm, itchy and scabby skin, wounds, and boils. Sufferers from gout, rheumatism, bladder stones, and asthma should avoid it.

Sorrel in large doses can damage kidney function. Over-indulgence in sorrel has caused poisoning in both humans and animals.

Here did she fall a tear,
here in this place
I'll set a bank of rue,
sour herb of grace.

WILLIAM SHAKESPEARE, *RICHARD II*

Rue

Rue is native to southern Europe, especially to areas around the Mediterranean. The Romans brought it to Britain, from where it was taken to the United States and Australia by settlers and explorers.

In the Renaissance, painters, engravers, and jewelers resorted to rue to improve both their eyesight and their creative inner vision; Leonardo da Vinci and Michelangelo are reputed to have used it. Rue had a strong reputation as a magical herb and was thought to grant second sight; it was also believed to be powerful against witches and their spells, and was hung above doors to ward off evil. Rue's reputation as a force for good led to its being adopted by Catholics as part of the Sunday ceremony of High Mass, when the priest used rue plants to sprinkle holy water on the congregation; it is from this usage that rue's common name of "herb of grace" derives.

Rue was thought equally effective as a safeguard against disease. It may have been an ingredient of a legendary "four thieves' vinegar" supposedly devised by a group of criminals to protect themselves from catching bubonic plague as they stole from corpses, and which subsequently passed into general use; some sources hold that rue was only used in a later version, known as "seven thieves' vinegar."

- Rue seeds are very fine; sow them into plug trays and cover lightly. They require a bottom heat of 68°F (20°C). Watch for damping off from over-watering.
- Rue is a hardy evergreen perennial. Take cuttings from new shoots in spring and early summer.
- Cut back plants after flowering to keep them in good shape.
- Rue prefers full sun and well-drained but poor soil.
- Growing rue in a container reduces any risk to the skin.

Rue is antispasmodic and a vermifuge (used to expel worms from the gut). It is one of the oldest known medicinal herbs. It was thought to ward off contagious disease and deter fleas and other biting insects. Herbalists may use it to treat hypertension (high blood pressure), but only under qualified supervision.

Rue is still used by some people, albeit sparingly, in egg, cheese, and fish dishes. Historically, it was added to sharp, acidic fruits such as plums, damsons, and gooseberries. Rue is sometimes used as a hedging herb for aromatic gardens, although it is not compatible with sage or basil, nor with cabbage.

The extreme bitterness of rue can lead to violent stomach upset in susceptible individuals. Rue should be avoided during pregnancy, because it causes the uterus to contract, and also while breast-feeding.

Nicholas Culpeper wrote that "the [rue] seed thereof taken in wine is an antidote against all dangerous medicines or deadly poisons." John Gerard advised its use against snake bites and sickness caused by eating poisonous fungi.

Handling common rue, or herbal preparations derived from it, can cause severe phytophotodermatitis, or burnlike blisters on the skin. Rain or full sun often are likely to strengthen the effect on the skin.

Common Sage

The genus name *Salvia* is derived from the Latin *salvere*, "to be well," a testimony to sage's historical reputation as a medicinal herb with health-giving properties. Indeed, in Classical times sage was known as *Salvia salvatrix* ("sage the savior") because it was believed to promote longevity. People aspiring to immortality, or just wishing for a longer life, would eat it every day. A popular rhyme encapsulated that aspiration: "He who would live for aye / Must eat sage in May."

A native of mountain slopes of southern Europe, the genus, especially the medicinal species *S. officinalis*, was taken much farther afield after it was identified as an effective remedy for numerous complaints. European monastic orders adopted it, planting it extensively in monastery gardens. The monks found that honey bees feeding on sage flowers alone produced a special honey that was itself considered to have curative properties. *S. officinalis* was introduced to North America by European settlers in the seventeenth century.

In tradition, a strong woman rules a household where the sage bush thrives. Curiously, she must not plant it herself, because doing so will bring bad luck. The sage must be planted by a stranger, and, even then, to avoid bad luck it must share its bed with other plants.

- Common sage grows from seed, cuttings, and layering.
- The pungent, hardy, evergreen perennial thrives in well-drained soil and full sun. It has gray–green, velvety leaves and mauve–blue-lipped flowers in summer.
- Cut back after flowering to leave up to 3 inches (8 cm) of green growth.
- Use as a companion for carrots and brassicas. Sage deters carrot fly and cabbage moth, and attracts bees and other pollinators.

Sage has antibacterial, antiseptic, and digestive properties. A decoction of sage may be used as a gargle for sore throats and gums. A fresh leaf of sage may be rubbed onto teeth to cleanse and protect them. Applied morning and night, sage oil relieves muscular pains and gives suppleness to joints.

Native North Americans would burn "smudge sticks" or bundles of white sage (*Salvia apiana*) in purification rituals. If that species was unavailable, common sage could be used instead.

In the Middle Ages, sage leaves burned on hot embers or boiled in an open saucepan would act as a natural aromatic disinfectant to cleanse a room after illness.

Sage is widely used in Italian and Middle Eastern dishes. Add fresh or dried leaves to risotto, pork, venison, duck, poultry, oily fish, potatoes, carrots, beans, eggplant, and tomato-based sauces. Dried sage has a scent of camphor; use it sparingly to bring out a dish's flavors but without itself dominating. Add sage flowers to salads or use as a garnish. Sage makes a delicious savory herb jelly for cold meat pies, cooked meats, and cheese. Sage may also be consumed as a refreshing beverage, or tea.

Be cautious when drinking sage tea. Consuming too much over a period of time can cause a toxin build-up.

It serves the doctor, the cook, the kitchen and the cellar, the poor and the rich.

Salvia sclarea

Clary Sage

Clary sage is native to Syria and southern Europe and is naturalized throughout Europe. It was cultivated in Britain from the mid-sixteenth century and is mentioned by John Gerard in 1597 as growing wild in "Holborne" and Chelsea in London.

The common name of "clary," coming from the Latin *clarus*, meaning "clear," derives from the herb's use as an eyewash; also, the seed has a mucilaginous quality that helps in removing foreign bodies from the eye.

The Germans infused the strongly aromatic herb in light white wine with elderflower to make a more powerful wine, a Rhenish imitation of muscatel. They called the herb *Muskateller Salbei*, which accounts for the English common name of "muscatel sage." Elsewhere in the drinks industry, clary sage is used as a flavoring and source of a lavender-like aroma in wines, vermouths, liqueurs, and even beers.

Nicholas Culpeper recommends clary sage "for swellings, tumors, and for drawing splinters," and "powder of the dried roots taken as snuff to relieve headaches." He also offers a culinary tip: "The fresh leaves, fried in butter, first dipped in a batter of flour, eggs, and a little milke, serve as a dish to the table that is not unpleasant to any."

In Jamaica, a decoction of clary sage leaves boiled in coconut oil was used by the native people to treat scorpion stings.

- Clary sage may be grown from seed sown in plug trays in early spring.
- Plant the seedlings outside when roots have developed.
- The plant likes a well-drained soil and a site in full sun. It will easily rot and die if its situation is too wet.
- The attractive aromatic biennial begins as a rosette in its first year, only producing flower stalks in its second year. It usually dies after that, though it may persist for one or two years more.
- A self-seeding plant, it is very suitable for borders.

Young clary sage leaves and flowers are added to salads. The young tops are used in soups and to flavor jellies and jams. Battered leaves and flowers are eaten as sweet fritters.

Clary sage is astringent, carminative, and antispasmodic. It is an emmenagogue (it promotes menstrual discharge), and treats menopausal symptoms such as hot flashes, palpitations, and irritability. It calms digestive upsets and is used to bring vomiting under control. Clary sage washes are used for leg ulcers, skin eruptions, cuts, and abrasions. In aromatherapy, clary sage is used to relieve depression, stress, insomnia, and tension.

The oil of clary sage is used as a fixative in soaps, cosmetics, and perfumes. It inhibits perspiration, reduces body odors, dispels excess oil in the hair, and eliminates dandruff.

Clary sage is an excellent companion for beans, cabbage, and carrots. The strong, unpleasant odor of the flowers may repel pests. Bees, however, are very attracted to the plant.

Do not use during pregnancy. Clary sage oil may be used to induce labor at term, but only under medical supervision because it can cause strong contractions, resulting in foetal distress.

And let the stinking elder, grief, untwine His perishing root, with the increasing vine!

WILLIAM SHAKESPEARE,
CYMBELINE ACT IV, SCENE 2

Common Elder

Common elder is native to Europe, North Africa, and West Asia. In North America, the species *Sambucus canadensis* more or less mirrors *S. nigra* in appearance and habit. Both usually grow wild in woods and along roadsides, in wastelands, and derelict areas, preferring rich soils where they can self-set freely and multiply. Flat, creamy, lacy blossoms are followed by fruits that are deep burgundy, almost black, in color.

Elder has many magical associations with witches, warlocks, sprites, and fairies. Depending on the culture, it was seen to bring doom and death, or promise life and regeneration. It was used as protective magic, and was grown near cottages to protect the inhabitants from lightning and witches. It was also believed that a healthy and vigorous elder tree offered evidence that the people living in its vicinity were happy. In Danish folklore, the Elder Mother, or *Hyldemoer*, a guardian spirit, haunted anyone who cut down a tree. It was considered unwise to lay a baby in a cradle made out of elder because the fairies would spirit it away. In England, hearse drivers would carry whips made of elder to protect themselves from malevolent spirits of the dead.

In rural Italy, people still make a musical pipe from elder, called a *sampogna*; it is related to the South American *zampoña*, or pan flute.

- Hardy perennial deciduous tree/shrub.
- Elder grows most easily from cuttings. A young twig broken off and pushed into moisture-retentive soil will almost certainly take root.
- Alternatively, take a handful of cuttings and, using a deep pot, fill with compost to just below the rim. Push the cuttings into the compost around the edge of the pot, leaving a space of at least 2 inches (5 cm). Keep the cuttings moist and in a shady spot until they have rooted.

Elder flowers, leaves, and berries are all used medicinally. A wash cools and soothes sunburn, sore eyes, and conjunctivitis. The flowers make a healing bath for bruises, stiff and sore muscles, sprains, and rheumatism, and a salve for burns and wounds. A tincture of the flowers promotes sweating in colds, fever, and influenza. The leaves repel insects such as gnats. A decoction of the berries, sometimes used with honey, may be used for coughs, colds, and catarrh, and as a gargle for sore throats.

Elder was once an important dye plant. A black dye is obtained from the bark, a green one from the leaves, and a purple one from the berries, all dependent on the mordant being used.

Elder flowers are commonly added to cold creams and hand lotions as a skin tonic and softener, and to treat acne, blemishes, rashes, freckles, and wrinkles. Elder flower is also used as a rinse to moisturize dry hair.

The flowers are made into cordials, syrups, jams, jellies, sorbets, ice cream, teas, wines, and vinegars. The cooked berries go into cold soups, sorbets, wild summer desserts, jellies, chutneys, sauces, and wines. Elder root can be made into a coulis. Young elder shoots are also eaten like one would eat asparagus.

Elder bark is purgative. Do not take it during pregnancy and while breast-feeding.

Sanguisorba minor

Salad Burnet

Native to central and southern Europe and as far north as southern Norway, salad burnet is also found in Asia, Armenia, and Iran. It was taken by the English pilgrims to New England, and has since become naturalized throughout North America. Green all winter, it grows wild in meadows, on the edges of woodlands, and by the sides of roads. It prefers chalk and limestone habitats and thrives on outcrops of chalk and rock. Farmers grow it as a highly nutritious fodder plant.

Interestingly, the flower heads are hermaphrodite, with female flowers on the top, and male flowers at the bottom.

The genus name comes from the Latin *sanguis sorbere*, meaning "to absorb blood," a reference to the herb's traditional use as a wound herb, applied directly to the injured area to staunch bleeding. The common name dates from the medieval period, when it was grown specifically in herb gardens for use in salads. The "burnet" part of the name refers to the brown coloring of the outer leaves, from the Old French, *burnete*.

- Salad burnet is best started off by seed in plug trays. Plant seedlings outside when roots are established.
- Likes a sunny position.
- The hardy, perennial, evergreen plants can be divided in the fall.
- Salad burnet is deep rooted and drought resistant.
- Snip off the flowers to increase the leaf crop.
- Plant near mint and thyme.
- Makes a good edging for an herb garden.
- Can be grown by the sea because it is salt tolerant.

Salad burnet is a good companion herb for thyme, and, unusually, it is not overwhelmed by mint.

Astringent, diuretic, and tonic, salad burnet was traditionally a cooling herb used to promote perspiration. It is recommended for gout and rheumatism. Culpeper and Gerard both refer to its being steeped in red wine for use.

The cooling, cucumber-like character of the leaves makes salad burnet an excellent herb for inclusion in leaf salads. It also suits Pimm's cocktails, long gin concoctions (the leaves frozen in ice cubes), and fresh berry-fruit salads. It goes with cream cheese and cottage cheese, vinegars and butters, and is delicious in vegetable fritters.

Cotton Lavender

Cotton lavender is native to dry hillsides of southern France and the Mediterranean. For centuries Arabic cultures used the herb medicinally in a lotion to soothe sore eyes. The Greeks called it *abrontonan* and the Romans *habrotanum*, both of which are references to its treelike appearance, each stem being like a trunk with branches. The herb is widely cultivated worldwide, notably in Australia and in drier, warmer areas of North America.

From medieval times in England the dried herb has been prominent in the home as an insect and moth repellent. Bunches were hung in rooms and closets, and dried leaves and flowers were collected in sachets and placed between folded linens to keep them free of insect pests. Stems were strewn on the floor for the same reason.

Nicholas Culpeper refers to the plant as a remedy for intestinal worms and skin irritations, saying also that it "resists poison, putrefaction, and heals the biting of venomous snakes."

Cotton lavender first became popular in Britain in the sixteenth century when French Huguenot gardeners used it in the "knot gardens" that became popular with Elizabethans of wealth. It is an herb that lends itself well to being clipped and shaped for hedging and edging.

- To propagate cotton lavender, take soft stem cuttings in the spring, or semi-ripe cuttings from non-flowering stems late in summer or early fall.
- This hardy, aromatic, evergreen perennial likes a well-drained soil and a position in full sun.
- Mass planting of cuttings is the most economical way of establishing a hedge.
- Cut back after flowering, or cut and shape in late spring.
- Cotton lavender loses color if the soil is too rich.

Antispasmodic and anti-inflammatory, cotton lavender was formerly used to treat insect bites and expel worms and parasites in children. It was also used as a compress, applied to wounds to promote healing and formation of scar tissue.

Layers of leaves used to be laid beneath woolen carpets as a moth repellent. Plants were also dried for potpourri.

Cotton lavender is of great benefit in a rose garden. Planted under or around rose bushes, it keeps moths and other insects from colonizing the plants.

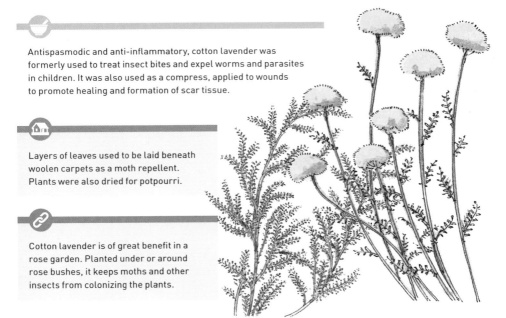

Soapwort

Soapwort has beautiful, pink-and-white, single or double flowers that are sweet-scented and especially aromatic on hot summer evenings. A native of central and southern Europe, the plant is thought to have been introduced to Britain in the Middle Ages; it is widespread in Asia and North America.

As the name suggests, soapwort has been valued since antiquity for its ability to produce a naturally cleansing soapy liquid. The Romans, who enjoyed washing and bathing, cultivated soapwort for this purpose, and plants may still be discovered growing wild on the sites of Roman bathhouses. In the Middle Ages, soapwort was known as "fuller's herb" because it was used by fullers, or cloth makers, to wash wool. In Switzerland, sheep were traditionally dipped in a wash of soapwort to clean their fleeces before shearing.

European immigrants in North America used soapwort to treat rash caused by poison ivy. Native Americans, such as the Cherokee, used it as a soap and poultice for boils. The Mahuna also used it for pain relief and as a poultice for application to the spleen; they used juice from the roots as a hair tonic and shampoo.

With its constituent saponins, soapwort is considered a useful precursor for other crops because it helps to condition the soil.

- Soapwort can be grown from seed but performance is erratic and stratification is needed.
- It is best grown from cuttings of the non-flowering shoots, taken in late spring, or division in the fall.
- The hardy, herbaceous perennial needs poor soil to check its rampant habit, or it can be invasive. It flowers from June to September.
- Do not plant around fish ponds because the saponin-rich rhizomes are poisonous.

Steeping or simmering the leaves and rhizomes produces a liquid soap that is suitable for cleaning natural fabrics with natural dyes, and hand-woven, naturally dyed woolen carpets.

Soapwort is purgative and anti-inflammatory. A decoction is used externally as a wash to treat psoriasis, acne, and eczema. It was formerly used for gout, rheumatism, and gonorrhea.

Soapwort is a mild poison and is not recommended for any internal use.

It raises a lather-like soap, which easily washes greasy spots out of clothes.

NICHOLAS CULPEPER,
THE COMPLETE HERBAL (1653)

Satureja hortensis

Summer Savory

A plant with an upright, bushy habit, summer savory is native to the Mediterranean regions, North Africa, and Southwest Asia. It was introduced to India, South Africa, and North America, where settlers planted it in the gardens of New England.

For more than 2,000 years, summer savory has been grown extensively both as an herb and for its stimulating properties as a medicine. The Egyptians put those properties to use by adding powdered savory to their love potions, and the Romans, too, grew the plant as an aphrodisiac. Indeed, the genus name *Satureja* is a reference to the lecherous satyrs of mythology, who lived in meadows of summer savory and were made passionate by eating the herb.

The Romans, who introduced the herb to Northern Europe, would use it in the same way as a mint sauce, adding it to numerous meat and egg dishes. Summer savory has excellent digestive properties and calms stomach cramps and gas; it is also mildly purging. It also has excellent antiseptic properties and was used by the Romans as a cleansing and strewing herb.

The flavor is hot, pungent, and peppery, like a strong thyme or mint with pine. Summer savory is known as the "bean herb" because it is widely used with broad beans, runner beans, and French beans in bean casseroles and soups. It also goes with pork and game (use sparingly) and is an ingredient of salami. It is found in *Herbes de Provence* with rosemary, oregano, and thyme, and used in vinegars, oils, marinades, savory jellies, and sweet, grape-juice–based jellies.

Summer savory is used as a digestion aid and diuretic. Its antiseptic properties may enable intestinal flora to revive and fight off bacterial infections. Use as a gargle for sore throats and gum or mouth infections.

Beside beans or onions, summer savory keeps beetles at bay and promotes vigorous and healthy growth. Allow some plants to flower for the benefit of bees.

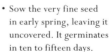

- Sow the very fine seed in early spring, leaving it uncovered. It germinates in ten to fifteen days.
- Harden off this half-hardy annual, then plant outside when no threat of frost.
- Plant in blocks or rows 6 inches (15 cm) apart. Summer savory likes a well-drained position in full sun.
- Cut the flowers when they appear or the flavor of the leaves will diminish.
- Harvest whole plants for drying before they flower.

Virginian Skullcap

Virginian skullcap is, as its name suggests, a native of eastern North America. In the wild it is found growing in damp areas near water courses. The genus name *Scutellaria* comes from the Latin *scutella*, which means "little dish" and refers to the shape of the calyx at the base of the flower. The plant has numerous other common names—blue skullcap, mad dog skullcap, Quaker bonnet, and helmet flower—and these refer to the resemblance of the calyx to the skullcaps worn in medieval times.

The name "mad dog skullcap" alludes to the plant's reputation for curing hydrophobia, a symptom commonly associated with rabies, the viral infection that targets the brain and nervous system and turns infected dogs dangerously mad. The efficacy of Virginian skullcap as a treatment for rabies was first asserted by Dr. Van Derveer of New Jersey in 1772.

There are around ninety species in the *Scutellaria* genus, of which *S. lateriflora,* so named because its flowers grow in one-sided racemes, is considered the most potent and effective medicinal plant. It has been claimed to be the finest nervine ever discovered.

Native Americans used the herb to promote menstruation, and some tribes used it in ceremonies to introduce young girls into womanhood.

- Grow from seed sown in the fall or spring.
- May be grown by division of mature plants in the spring.
- Virginian skullcap prefers damp conditions and fertile soil.
- This hardy perennial is pretty enough to grace a flower border, but be warned that it can become invasive if left unchecked.
- Cutting back after flowering will help to counter its self-seeding habit.
- Harvest the aerial parts of the plant for use in infusions, extracts, and tinctures.

Virginian skullcap is antispasmodic, digestive, nervine, and sedative. Taken as an infusion, it soothes and calms nervous anxiety, stress disorders, hysteria, and tension. It helps to lift depression and is used for insomnia, neuralgia, and sciatica. It eases menstrual problems, including premenstrual syndrome (PMS).

Do not take Virginian skullcap internally during pregnancy because it can cause miscarriage, nor during breast-feeding. Avoid it also when using tranquilizers or sedatives. Virginian skullcap can cause confusion and twitching in some people and should be taken only under the supervision of a qualified herbalist.

Modern-day witches tend to prefer Virginian skullcap or wild lettuce (*Lactuca virosa*) for use as a "flying ointment" because herbs traditionally used, such as belladonna and henbane, are very toxic.

Virginian skullcap is used for treating anxiety because, unlike valerian (*Valeriana officinalis*), kava (*Piper methysticum*), and hops (*Humulus lupulus*), it causes minimal drowsiness.

Virginian skullcap was used as a ceremonial plant by some Native American tribes, who would smoke it as tobacco to induce visions. However, an overdose can cause giddiness and stupor.

. . . and that plant they keep against evil atop their house.

JACOB GRIMM, QUOTING AN ITINERANT MUSICIAN

Houseleek

The houseleek belongs to a genus of over 500 hardy succulents but seems to be the only one with a history of medicinal practice. It is native to mountain ranges of central and southern Europe and the Greek islands, and was introduced to other parts of the world, including North America, by early settlers.

The genus name *Sempervivum,* deriving from the Latin *semper,* meaning "always," and *vivo,* "alive," refers to the plant's longevity. The species name *tectorum* means "of the roof," and records show that the herb has been grown on roofs for 2,000 years, especially on tiled roofs. Emperor Charlemagne ordered his subjects to grow houseleeks on their roofs to ward off lightning, and this superstition persists in many parts of the world. In medieval Britain, houseleeks were grown on the roofs of cottages and outhouses, although there it was believed that they had magical powers to ward off witches.

Another common name, "Jupiter's eye," refers to the Roman god Jupiter and the watchful protection he gave man against lightning and fire. The Romans grew the plant in urns around their courtyards. In the medieval period, however, the flowers were thought unlucky because they resembled staring eyes; as a result, many would be cut off.

The name "hen and chicks" refers to how the plants grow a large center crown with many smaller plantlets clustered around it.

- Hardy evergreen perennial.
- Seeds are unreliable because they hybridize easily, so propagate by offsets that will keep the plant "true."
- Each little cluster, when separated from the parent plant, will have its own roots, which makes transplanting an easy task.
- Use a gritty compost and small pots—they like a tight space. Transfer in late summer or early fall.
- Plant in a well-drained rock garden, shallow trough, or dish-shaped pot, in full sun.
- Do not over-water. The plants are drought resistant.

A few plants set on a roof spread quickly by sending out offsets, and these may be detached and used to establish new colonies. In this way, houseleeks develop into a covering that preserves the roof. Thatched roofs are particularly benefited.

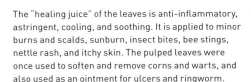

The "healing juice" of the leaves is anti-inflammatory, astringent, cooling, and soothing. It is applied to minor burns and scalds, sunburn, insect bites, bee stings, nettle rash, and itchy skin. The pulped leaves were once used to soften and remove corns and warts, and also used as an ointment for ulcers and ringworm.

Crushed leaves may be laid against the temples to alleviate headache, and the leaves may be chewed to ease toothache. A gargle made with pulped leaves is good for a sore throat.

Houseleek may be used as a facial steam to nourish and heal the skin, or applied directly to red, inflamed patches. If preferred, an ointment may be made with crushed leaves and lard.

Young houseleek shoots and leaves are crunchy and cooling with a cucumber flavor and texture. Leaves can be pressed for their juice, but drink it with caution because it can be purging.

Sesame

Thought to be native to Africa, *Sesamun indicum*, commonly known as sesame or gingelli, has been cultivated in parts of Asia, Africa, China, and India for thousands of years. Ancient papyrus records mention its importance and use in Egypt 5,000 years ago. It was introduced to Mesopotamia in around 2,500 BCE, where the Assyrians used it as a food and as a medication. There are references to sesame being grown in Italy during the time of Pliny (23–79 CE), and Dioscorides wrote of the Roman custom of sprinkling sesame seeds on Sicilian breads.

Sesame is one of the first plants from which humankind extracted oil for use as a condiment. The oil has similar properties to olive oil, and has been produced domestically in the Middle East for centuries. Sesame offers an economic lifeline to developing countries that encounter persistent drought because it grows where other crops fail. The plants regenerate efficiently because of the way the seedpods swell and burst open, scattering seed onto any receptive ground. In *Ali Baba and the Forty Thieves*, Ali Baba orders the "cave of treasure" to open with the words "open sesame," the command being an allusion to the explosive opening of a sesame seedpod.

For Hindus, sesame oil is the correct oil for burning in the lamps used to light sacred shrines, and for anointing effigies of deities.

- Propagate by seed sown in spring or fall; simply take a few from the herb cupboard and sow them.
- Water the seeds well at first to encourage germination, and then only as needed.
- Sesame plants prefer a well-drained sandy soil in full sun.
- In a temperate climate, sesame needs a relatively dry season to do well and produce seeds.
- Harvest the seeds when ripe and store unhulled in an airtight container in a cool, dry, dark place.

Sesame is known as a "food medicine," being anti-inflammatory, antispasmodic, tonic, and laxative. Rich in vitamins, it is given as a convalescence tonic, and its calcium content strengthens bones and teeth. It is used to treat constipation, osteoporosis, stiff joints, dry coughs, tinnitus, poor vision, and liver and kidney disorders. It may be taken internally to prevent hair loss and graying.

Sesame products carry undoubted health benefits but the seeds and oil are high in calories. Use them in moderation if weight gain is to be avoided.

In India, sesame oil is used neat as a massage or cosmetic carrier oil because it penetrates the skin easily. It also contains vitamin E and beneficial acids, and the antioxidant sesamol, which helps to prevent skin ageing.

The seeds are used whole, pressed, and as a paste and oil. In Jordan, Syria, and Lebanon, the *zahtar* spice mix includes dried untoasted sesame seeds, thyme, and sumac; in Egypt, the *dukka* spice mix has sesame with coriander and cumin. In the Middle East and East Asia, seeds are ground into tahini, used in hummus, and in halvah, a treat made with honey. Seeds are sprinkled into or over many Japanese, Chinese, Korean, and Burmese dishes. Sesame oil is good for stir-fries, salad dressings, roasting, and in marinades.

Love is the talisman of human weal and woe—the Open Sesame to every soul.

ELIZABETH CADY STANTON,
SUFFRAGIST AND FEMINIST

My opinion is that this is the best remedy that grows against all melancholy diseases.

NICHOLAS CULPEPER, *THE COMPLETE HERBAL* (1653)

Milk Thistle

Milk thistle is native to southwest Europe and was introduced to Belgium, Holland, Denmark, and the British Isles, where it grows wild in coastal areas, on hedgebanks, and wastelands. It is naturalized on dry, rocky soils in North America.

The Latin name translates as "annual herb of Mary," a reference to the Virgin Mary. The common name also refers to her, albeit obliquely, in that the beautiful white veining of the first-year leaves was observed as having come from her mother's milk. The name, and the Doctrine of Signatures, led to the belief that the plant was a galactagogue (able to increase breast milk), and thus it was grown in monastic gardens from early medieval times to be given to nursing mothers so that they could also breast-feed other women's babies. However, modern research had cast doubt on that claim.

More reliably, the herb was found to protect the liver. Pliny the Elder refers to its use for "carrying off bile" and Dioscorides saw it given to those "bitten of serpents." It is now known to be effective in the treatment of all types of hepatitis, gallstones, and cirrhosis, as well as poisoning from ingestion of the death cap mushroom, drugs, and alcohol. Silymarin extract from the plant protects the liver by blocking absorption of these powerful toxins. An even more amazing attribute of silymarin is that it stimulates the production of new liver cells.

- Grow from seed sown in seed trays or plugs, or straight into a prepared site in the ground in the summer or early fall.
- Milk thistle likes a well-drained site in full sun.
- The foliage of this hardy biennial is at its best before the plant flowers.
- Removing the flowers will prolong the show of white-veined foliage.
- The plant self-seeds freely and can be invasive. It is subject to environmental control in some countries.
- Seeds can be harvested for a few weeks.

Milk thistle is antioxidant, anti-inflammatory, and mildly laxative. Because its main active ingredient, silymarin, is not easily obtained from infusion in water, it is usually prepared in the form of a tincture. As a liver tonic, it removes toxins, regenerates cells, and increases function, and so is used to minimize the effects of chemotherapy on the liver. The herb eases gallbladder problems by increasing bile production; another benefit is that it cleanses the blood.

Young leaves with their prickles removed, and stalks peeled and sliced, may be eaten in salads. The roots are similar to salsify and are eaten as a cooked vegetable. The flower heads make a good, if smaller, substitute for artichoke. The nutrient-rich edible sprouts (harvested as the seed germinates) are high in antioxidants. Juice extracted from the herb is used to flavor vodka cocktails.

Used as an attractive edging plant, milk thistle will deter cats from venturing onto newly planted herb borders. The seed heads will long remain in place and provide winter feed for birds.

Extracts from milk thistle may act like estrogen and are best avoided by women with hormone-sensitive conditions such as cancers of the breast, uterus, or ovaries, or endometriosis.

The Romans left but the Alexanders stayed behind . . .

JOHN WRIGHT, IN "HOW TO MAKE GIN ALEXANDERS," IN THE *GUARDIAN*

Smyrnium olusatrum

Alexanders

The odd-sounding names of "Alexanders" or "alisanders" are shortenings of what the Romans called "the herb of Alexander the Great." By the same token, *Smyrnium olusatrum* is known as Alexandrian parsley, and, in a reference to another of its flavors, Roman celery. It was introduced into Britain by the Romans and for centuries was used for culinary purposes until largely replaced by true celery (*Apium graveolens* and its cultivated varieties).

The plant is native to western Europe and the Mediterranean region. A prolific self-seeder, it is now found naturalized, especially in coastal areas and lands near old monastery gardens and medieval manor houses where extensive vegetable gardens would once have existed. At that time it was grown as an herb cum vegetable for its flavor, a blend of parsley, celery, and chervil. Maud Grieve, author of *A Modern Herbal* (1931), lists it as "black lovage," her choice of color being a reference to the jet-black seeds. She writes: "The plant received from the old herbalists the name of 'Black Pot-herb,' the specific name signifying the same (*olus*, a pot herb, and *atrum*, black). Resembles both lovage and angelica and by the inexperienced can be wrongly identified."

The herb is also known as horse parsley because the animals are known to favor it while grazing.

- Like many of this umbel family, Alexanders is best cultivated from seed, sown fresh and ripe in the late summer or early fall.
- Sow in seed trays and place outside with glass over them to prevent rodents from eating the aromatic seeds.
- Once established in a garden, this hardy biennial or perennial will self-seed if left alone.
- Pick leaves throughout the summer, stems when young, shoots in spring, and roots in the late summer of the plant's second year.

Alexanders benefits other food crops by being very attractive to pollinating insects. In 1977 an English observer listed 137 different species of insects drawn to *S. olusatrum* flowers.

Rich in vitamin C, Alexanders is antiscorbutic (it prevents scurvy), and it was once officially prescribed for that purpose. It is mildly diuretic and the bitter root was decocted for the treatment of urinary complaints. The roots are also digestive, an appetite stimulant, and stomachic, and are therefore recommended as an aid to digestion. The juice may be placed directly on cuts and wounds to assist healing, and also used to treat asthma.

As an herb, Alexanders was widely used in soups, casseroles, and stews. The dried seeds—which contain an essential oil, cuminal, not unlike cumin or myrrh—can be ground like peppercorns and used to season foods. Flower buds may be eaten raw in salads, steamed, or pickled. The leaves, especially in spring, offer added nutrition in salads. The stems are steamed or braised like asparagus. In the fall, the roots may be eaten like parsnips.

Alexanders probably lost popularity in relation to celery as part of a general Western trend away from pungent, bitter foods and toward more tender, sweeter alternatives.

Goldenrod

Goldenrod is a European native that also grows wild in North Africa and parts of Asia. North America has a number of its own *Solidago* species, more commonly called Aaron's rod, all with similar properties to *S. virgaurea*, and these were used medicinally by Native Americans.

The genus name is from the Latin *solidare*, meaning "to make whole or strengthen," a reference to the herb's healing properties. Nicholas Culpeper refers to goldenrod as "a Sovereign wound herb inferior to none, both for inward and outward use."

In the Middle Ages, goldenrod was promoted by the Arabs, who exported it from the Middle East to other countries. Its virtues were extolled by the Saracens, who used it to treat their wounds after battling with the Crusaders. In the fifteenth and sixteenth centuries it was used by the Italians, who called it *erba pagna*, and the Germans, who called it *Consolida Saracenia*—further references to its use for healing wounds. It was imported into Britain and commanded high prices until it was discovered growing wild in London, in Hampstead Wood.

In 1929, American inventor Thomas Edison experimented with goldenrod's ability to produce rubber. By breeding the herb extensively he increased the latex content by 12 percent, and the rubber proved to be resilient and long lasting. A Model T Ford motorcar, given to him by his friend Henry Ford, had tires made from goldenrod rubber.

- Sow the fine seed into plug trays. Germination should occur within fourteen and twenty-one days.
- Plant the seedlings in the garden after hardening off.
- The hardy perennial can be divided in either spring or fall.
- Goldenrod likes sun and a well-drained soil.
- The plant has a common name of "farewell summer" because the sprays of bright yellow flowers appear in late summer.
- The plant can be invasive. As a precaution, remove the spent flowering tops.

Goldenrod is astringent, anti-inflammatory, expectorant, vulnerary, and diuretic. It is used to treat cystitis and urethritis, as well as arthritis and rheumatism. It prevents kidney stones. Homeopaths have used it since 1902 for respiratory, urinary, and uterine problems. A healing compress may be applied to a fresh wound or inflamed skin. Tea made from dried leaves and flowers is aromatic and diuretic, and can be used as a mouthwash for loose teeth.

Goldenrod is loved both by bees and predatory insects such as lacewings, ladybugs, and hoverflies. Growing it in the garden increases the numbers of such predators, which in turn rid the garden of pests such as aphids.

Goldenrod was used as a divining plant. Held upright in the hand, it was believed to nod in the direction of lost objects or buried treasure.

The leaves and flowers of goldenrod both yield a strong yellow dye of reliable fastness, although only craft dyers tend to use it.

All the aerial parts of the plant are edible. The flowers make attractive garnishes on salads. The leaves can be cooked like spinach, added to soups, stews, or casseroles, or blanched and frozen for later use in stir fries.

Graceful, tossing
plume of glowing gold
. . . Leaning seaward,
lovely to behold . . .

CELIA THAXTER,
"SEASIDE GOLDENROD" (1874)

Stachys officinalis syn. *Betonica officinalis*

Wood Betony

Native to Europe, wood betony is found growing wild in Britain in woodland clearings and meadows. Its attributes as a magical plant were first noted by the Egyptians. The herb was particularly important to the Anglo-Saxons, who used it to treat "elf sickness" and ward off bad magic and enchantments wrought by witches. Deemed powerful against evil, it was planted in churchyards, hung in bunches around doorways, and worn as an amulet, "driving away devils and despair."

The genus name *Stachys* derives from the Greek *stachus*, a spike resembling an ear of wheat. Medieval herb gardens gave betony pride of place and it was held in high esteem by herbalists for the treatment of gout, fever, worms, chest and lung problems, liver and kidney disorders, and venomous serpent bites. The twelfth-century Welsh physicians of Myddfai wrote of the herb: "If boiled with leek seeds, it will cure the eye and brighten as well as strengthen." Romani people in Derbyshire, England, infused the leaves for stomach troubles; they also made ointment from lard and the juice of the fresh leaves to neutralize poison from stings and bites.

Nicholas Culpeper made many recommendations for wood betony, saying it could be "either green or dry, either the herb or roots or the flowers in broth, drink or meat or made into conserve, syrup, water, electuary or powder, as everyone may best frame themselves unto, or as the time and season requireth."

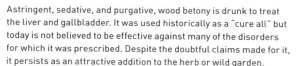

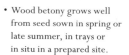

- Wood betony grows well from seed sown in spring or late summer, in trays or in situ in a prepared site.
- The roots may be divided in spring or fall.
- The hardy perennial prefers humus-rich soil in a sunny or half-shaded site.
- Usually a wildflower, it is also cultivated in herb gardens.
- Cut off flowering spikes to control its spread through prolific self-seeding.

Astringent, sedative, and purgative, wood betony is drunk to treat the liver and gallbladder. It was used historically as a "cure all" but today is not believed to be effective against many of the disorders for which it was prescribed. Despite the doubtful claims made for it, it persists as an attractive addition to the herb or wild garden.

Betony is used as a rinse to remove yellow streaks from gray hair, and as a hair brightener. It also helps to eliminate head lice and heal their effects on the scalp.

Taken internally, wood betony causes vomiting and diarrhea. The herb is thought to affect blood pressure; avoid it for at least two weeks before surgery.

Comfrey

Comfrey is native to Europe and temperate Asia. It grows wild in damp, shady places, river banks, brook sides, and bogs. It is thought to have been brought to Britain by returning Crusaders who had witnessed its healing properties. From there it was eventually carried to North America for use in herbal and medicinal gardens; escaping from these, it has since naturalized in the wild.

The genus name *Symphytum* derives from the Greek *symphysis*, the growing together of broken bones; *officinale*, as always, means that it is listed officially as a medicinal plant. In his *Materea Medica*, Dioscorides recommends comfrey for healing broken bones and wounds. The common name comfrey, from the Latin *confera*, "to knit together" has the same meaning, and other common names—knitbone, boneset, bruise wort—also refer to the herb's most valuable medicinal uses.

Comfrey contains allantoin, a cell proliferant, vitamin B$_{12}$, and as much protein as soy, but researchers discovered that it has potential to damage the liver. Tablets and tinctures for internal use are now banned in several countries, but external use of comfrey is still thought safe.

- Comfrey may be grown from seed, but it is much simpler and quicker to use root cuttings or divide roots in either spring or fall.
- This hardy herbaceous perennial prefers moist soil, in sun or shade.
- Comfrey can be rampant and take over very quickly.
- Cut back after flowering before the seeds start to spread.

Whole young leaves can be made into fritters and are great for making *pakoras*. Always use in moderation, and only the young leaves.

Comfrey flowers early in the year and its nectar is very attractive to bees. Rotted down, comfrey plants make a putrid-smelling liquid feed that is an excellent source of nitrogen and potassium. The feed may be dug into potato trenches or used for tomato nutrition.

Wear protective clothing when handling; comfrey can irritate the skin.

Comfrey leaves are more potent before the plant flowers. They make a compress for varicose veins, sprains, swellings, burns, wounds, cuts, bruises, and sores. A comfrey salve helps to heal bones, boils, hemorrhoids, and rheumatics.

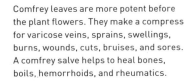

Mexican Marigold

This plant, native to Mexico, is also known as Mexican tarragon and Spanish tarragon, and, like French tarragon (*Artemisia dracunculus*, see p. 34), it is a member of the family Asteraceae. The plant is grown in North America as a tarragon substitute, especially in the south, where conditions are too hot for French tarragon to grow.

In the Mexico of the Aztecs the plant was especially used to flavor *chocólatl*, a foaming, savory, cocoa-based concoction that was reputedly hallucinogenic and anesthetic in its effects. Mexican marigold was also a ritual incense herb, used to celebrate the Day of the Dead and other festivals along with other species of its genus. Known in Mexico as "cloud plant," it was the source of a powder used by the Aztecs to calm the terror of those imminently to be offered as human sacrifices. Mexican marigold is a shamanic herb and is still used by shamans, especially as an aphrodisiac. It is hallucinogenic when smoked while partaking of liquors such as maize beer and cactus wine.

Mexican marigold is related to the French marigold (*Tagetes patula*) and African marigold (*T. erecta*), both of which are widely planted as ornamentals. All three should not be confused with the pot or English marigold (*Calendula officinalis*, see p.41), which is a distant relative.

- Seeds may be sown in early to late spring. They require a temperature of 20°C (68°F) to ensure germination.
- Sow seeds into plug trays to minimize later disturbance of their newly formed roots.
- Plant this tender perennial outside when the threat of frosts has passed.
- Mexican marigolds like full sun and a well-drained position.
- Leaves may be harvested as soon as they grow, either one at a time or as sections from the tips of branches.

Secretions from Mexican marigold roots repel some slugs, nematodes, and beetles from crops such as asparagus, but there is also an herbicidal effect on plants like beans and cabbage.

Dried Mexican marigold plants are repellent to insects that infest the home, and are especially effective against bean weevils. The dried herb is also burned indoors as incense to cleanse the air. The flowers are processed to obtain a dye that can range from orange to golden and pale yellow.

Mexican marigold tea can ease indigestion, nausea, and fevers, and is traditionally given to people who have been struck by lightning. The herb is used externally to treat eczema scorpion bites, and ticks.

Tagetes species can cause an allergic reaction and are best prescribed as an internal medicine by herbalists only.

Tamarind

Tamarind, an evergreen tropical tree thought to be native to Africa and Madagascar, now grows widely throughout the tropics. Traditionally the bark was used in African countries to treat and heal wounds. After being cultivated in India for many centuries, it was taken by the Spanish to the West Indies and Mexico in the seventeenth century. Records indicate that the tree was in cultivation in the Middle East from the fourth century BCE.

Tamarindus, from the Arabic Tamar-Hindi for "date of India," owes its name to the datelike pulp inside the pods. The Indian epic poem *Ramayana* relates how the leaves of the tamarind tree became serrated: Lord Rama's half-brother, Laxmana, had built a hut out of tamarind leaves as a home for the exiled Rama and himself. Unfortunately the leaves allowed no light to penetrate the hut, so Laxmana shot at them with his bow and arrows until they were shredded enough to let in light. The tropical leaves have been serrated ever since.

Tamarind's tart taste features in certain Asian curries, duck and pork dishes, satays, and fish dishes. Sweet tamarinds are eaten like dates. Fresh leaves are eaten as a vegetable. Roasted seeds are ground into flour for baking.

Flour from the ground seeds can cause contact dermatitis and asthma.

Tamarind is taken internally for fevers, jaundice, asthma, dysentery, and nausea in pregnancy. It is combined with senna as a laxative.

Tamarind wood is used as timber, firewood, and charcoal. The wood is traded in North America under the name of Madeira mahogany. A yellow dye may be obtained from the leaves.

- Tamarind may be grown from seed in slightly acid conditions in spring. Sow under glass an maintain at above 70°F (21°C).
- Cuttings may be grafted.
- This tender tree requires tropical conditions to thrive.
- Tamarind can be grown as a potted plant in a warm glasshouse or conservatory.
- The fruits may be picked when ripe, although they can stay on the tree for up to six months. Growers often compress them into blocks.

Alecost

Alecost, also called costmary, is a native of western Asia. It was grown in England for its range of uses for many centuries, before finding its way to America by means of the early settlers. It now grows wild in parts of America but has become very rare in Europe.

The common name "alecost" comes from the herb's use in preserving, clarifying, and flavoring ale ("cost" comes from the Greek *kostos*, meaning "spicy." In fact, in Britain the herb is also called "spicyale." The name "costmary" refers to the Virgin Mary and alludes obliquely to the herb's use as an aid to childbirth in medieval times.

A North American common name, "Bible leaf," refers to the leaves being used as Bible book markers, partly because they were thought to repel book worms; another explanation of the name is that they were chewed to allay boredom through long Christian sermons.

Alecost is closely related to the camphor plant (*Tanacetum balsamita* subsp. *balsametoides*), which is very similar in appearance but has a strong camphor scent. Alecost itself has a balsam-mint fragrance and a minty-lemony taste that makes it preferable as a culinary herb.

- Alecost is propagated by division of mature plants in either spring or fall.
- Place pieces of the creeping roots into appropriately sized pots using a good mixed potting compost.
- Both alecost and camphor are hardy herbaceous perennials that prefer a well-drained soil and a sunny site.
- Cut back both plants after flowering to encourage growth of additional and more potent leaves.

Alecost is added in small quantities to salads, soups, root vegetables, potatoes, peas, beans, poultry, and meat-based stuffings.

Historically alecost was used as a tea to ease childbirth. It is also an aid to digestion and a liver tonic. It helps to expel worms in children, and is used externally to relieve insect bites and stings.

An alecost rinse adds shine and luster to hair. In a bath it soothes and relaxes, and can alleviate symptoms of a cold.

Whole bunches of alecost or camphor plant are hung in bags in closets to deter moths; both are also used in sachets with lavender to protect and scent linen. In potpourri they are used to enhance other plant scents.

Alecost is not used in modern herbal medicine because other herbs are considered more potent.

Feverfew

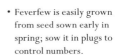

A pungent herb thought to be native to southeastern Europe, feverfew is prolifically self-seeding and has spread throughout Europe and both North and South America. As a garden escapee, it is found on wastelands, scrub, and rocky places. At one time its Latin name was *Chrysanthemum parthenium*, a reference to the shape of its leaves.

The common name "feverfew" is misleading because the herb's main medicinal strength actually lies in its ability to alleviate migraine—a power that has been confirmed by remarkable modern research results. Feverfew has been used for headaches from the first century AD; it was later used as a remedy for joint inflammation, arthritis, premenstrual syndrome, fevers, aches, and pains. Nicholas Culpeper wrote that "Venus commands this herb, and has commended it to succor her sisters," and his patients used it for menstrual problems.

At the time of the European plagues, feverfew was planted around homes to prevent illness and ward off evil spirits. In the sixteenth and seventeenth centuries it continued to be grown in botanical gardens for its healing properties. Today, many gardeners are deterred from growing it by its rampant self-seeding habit.

Feverfew should be avoided by anyone taking blood-thinning medication. It is a uterus stimulant, so avoid it if pregnant. It can cause mouth ulcers as a side effect.

As an insect repellent related to pyrethrum, it may be sprayed to deter mushroom fly, moths, aphids, and red spider mite—but it repels bees, too.

- Feverfew is easily grown from seed sown early in spring; sow it in plugs to control numbers.
- Germination should occur within two weeks.
- After it has rooted, plant outside in a sunny position. It will establish itself in any rocky crack or crevice.
- The hardy perennial is a prolific self-seeder; remove flowers before they set seed.
- To maximize the potency of the leaves, harvest them before the plant flowers.

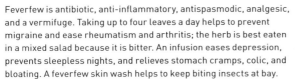

Feverfew is antibiotic, anti-inflammatory, antispasmodic, analgesic, and a vermifuge. Taking up to four leaves a day helps to prevent migraine and ease rheumatism and arthritis; the herb is best eaten in a mixed salad because it is bitter. An infusion eases depression, prevents sleepless nights, and relieves stomach cramps, colic, and bloating. A feverfew skin wash helps to keep biting insects at bay.

Tansy

A native herb of Europe and Asia, tansy is found naturalized through many parts of the Northern Hemisphere, including in North America. In his book, *Herbs and Earth* (1935), U.S. author Henry Beston wonders which plants should be considered for a patch of herb garden in the United States; his comment—"Here Tansy can go in if you want so familiar a weed at all"—testifies to its prolific presence in that country.

Ever since Zeus gave it to Ganymede to make him immortal, tansy has been carried and grown as a symbol for longevity. The herb was once called *athanasia*, Greek for immortality. That name, referring to the long-lasting flowers, is thought to be the source of the common name. The plant was at first dedicated to St. Athanasia, and later on to the Virgin Mary.

Tansy cake is an Easter tradition among the Christian clergy. Made from the leaves mixed and bound with egg, it was eaten to purify the body after Lent. Tansy was also used at funerals. Tansy wreaths were placed on coffins before burial, and coffins were packed with the herb to help preserve the corpse. Consequently, tansy became known as the "corpse herb," and it soon became synonymous with death.

Tansy acts as a deterrent for ants, flies, and mice. At one time, fresh meat was rubbed with tansy leaves to keep blowflies away.

In the past, tansy seeds were used to make a tonic tea. The leaves went into an enema for expelling worms, and were used externally as a compress to treat scabies and painful joints. Traditionally, tansy was used as an abortifacient.

Planting tansy by roses, squashes, and cucumbers is believed to invigorate them. It also attracts ladybugs, which will prey on aphids. Tansy is a valuable compost plant, rich in potassium and other minerals, but use only the leaves.

Overdose of tansy tea or oil can be fatal, and tansy oil is a banned substance in many countries. For the use of qualified herbalists only.

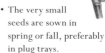

- The very small seeds are sown in spring or fall, preferably in plug trays.
- When the roots are clearly established, plant outside in moderately fertile soil, in a site with plenty of light.
- Take care where you plant tansy because it spreads easily, both by rhizomes and self-seeding.
- Divide mature clumps of this hardy perennial in spring.
- Cut back after flowering.

Taraxacum officinale

Dandelion

Taraxacum is a prolific wild herb genus; there are more than 1,000 recorded species of dandelion. It is first mentioned as a medicine in a Chinese record of 659 CE, although it is thought that Pliny advocated its uses at an earlier time. Arab physicians extolled its virtues in the eleventh century, and it appears in European medical texts from the fifteenth century onward.

In the thirteenth century, the Myddfai physicians in Wales administered dandelion to treat yellow jaundice. The yellow color of the flowers, according to the Doctrine of Signatures, suggested its use for that purpose. The milky juices in the plant reputedly "thwarted warts," and were also fed to young animals to nourish them. The common name "wet-the-bed," *pissenlit* in French, refers to the herb's potent diuretic properties.

A tradition of blowing the seed heads to tell the time is recognized in the common name of "fairy clocks." The flowers also have weather-vane abilities, fully extending in fine weather and closing up when atmospheric pressure decreases and rain is due.

Dandelion is a natural diuretic, laxative, and digestive, and is rich in potassium. It is a tonic herb for the kidneys and liver. Flowers may be boiled with honey as cough remedy. The milky latex removes warts.

Dandelions are beneficial in orchards because they release ethylene, a gas that encourages fruit setting and fruit ripening. Also, the roots attract earthworms.

> *Dandelion comes from the French "dent de lion," meaning lion's tooth.*
>
> PAUL SIMONS, THE *GUARDIAN*

Used in salads, young leaves are good with bacon and chicken. They are rich in vitamins. The roots may be roasted as coffee, and the flowers made into wine.

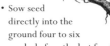

- Sow seed directly into the ground four to six weeks before the last frost.
- Once seedlings have sprouted, thin to 6–8 inches (15–20 cm) apart.
- The hardy perennial can be grown as an annual to achieve sweeter leaves.
- Dandelion has a long taproot and self-seeds easily. It is a very difficult weed to eradicate successfully.

Wall Germander

A small, shrubby, evergreen herb, wall germander is a member of the mint family that is native to mountainous regions of Europe and Southeast Asia. It thrives on dry, rocky, sunny slopes, and is thus ideally suited to the dry environment of garden walls, which, naturally heating up and cooling down with the movement of the sun, expose the plants to a wide temperature range. *Teucrium chamaedrys* is a notably hardy plant, able to withstand temperatures down to -18°F (-28°C).

The genus name comes from the name of a Trojan king, Teucer, a famed archer who supposedly used the herb as a medicine. The species name is derived from the Greek *chamai,* meaning "ground," and *drus* meaning "oak," a reference to the shape of the leaves. Ground oak is one of the common names of the plant.

Dioscorides recommended wall germander for coughs and asthma. The herb is reputed to have cured the Holy Roman Emperor Charles V (1500–58) of gout after he took a decoction of it for sixty days. Nicholas Culpeper remarked that it "strengthens the brain and apprehension exceedingly when weak, and relieves them when drooping" and that it is "good against all diseases of the brain, as continual head-ache, falling-sickness, melancholy, drowsiness and dullness of the spirits, convulsions and palsies."

- Propagate by seed sown in spring. Germination is erratic and can take many weeks.
- An easier option is to take softwood cuttings from new spring growth.
- The hardy, evergreen perennial likes free-draining soil in full sun. The plant is drought resistant.
- Cut back after flowering; more cuttings can be taken from the regrowth.
- If used for edging, set plants 12 inches (30 cm) apart.
- The plant is good for cascading over walls.

The herb remains important in the treatment of gout. It was traditionally used to improve digestion and increase appetite, and to treat gallbladder disorders and stomachache.

Wall germander is widely used in making alcoholic drinks with a bitter base, and that are digestive or appetite-promoting, such as vermouths, tonic wines, and certain liqueurs.

In the garden, low-growing wall germander is mainly used to clothe stonework. If a taller plant is required, a relative, hedge germander (*T. divaricatum*), is often used, trimmed as a dwarf hedge.

Wall germander is commonly planted to attract pollinating insects to food crops and ornamental planting. Bees love this plant and have been observed to ignore others wherever it is planted.

In the early 1990s, wall germander was marketed, both alone and with other herbal ingredients, as an aid to weight loss. It caused dozens of cases of toxicity and one fatality (see right).

Wall germander is rarely used in modern herbalism. Its use as a slimming aid has been banned in France because it causes hepatitis, liver and kidney damage, and is potentially lethal.

Thyme

The genus *Thymus* includes numerous species that are found around the world; the majority are native to the Mediterranean region. Thyme has been grown commercially and domestically for thousands of years. Strangely, in the ancient world its sweet aroma was associated with death, and so it was given a role in funerary rituals. It was thought that the souls of the dead resided in its flowers, and the scent of thyme has been reported on the air of several known haunted sites.

The Egyptians used thyme for embalming, and the Greeks used it for purification and consecration purposes, burning it like incense in their temples. Both the Greeks and Romans made oils from thyme for use in medicines, and also for aromatherapeutic massage and bathing. The Romans are credited for the cultivation and spread of thyme westward through Europe to the British Isles.

Thyme has long been associated with strength and courage. At the time of the Crusades, ladies would embroider thyme sprigs and bees onto their knights' clothing and pendants to make them brave in battle.

During the Great Plague, rat fleas infected by the bacterium *Yersinia pestis* caused the bubonic plague to spread. Thyme is insecticidal and was used as a strewing herb to deter the fleas from infesting homes; people also carried thyme inside their clothing as personal protection.

- Grow only common thyme (*T. vulgaris*) and wild thyme (*T. serpyllum*) from seed. Other thymes cannot be guaranteed to "come true."
- Grow from softwood cuttings of new growth in spring or early summer.
- Divide creeping thymes from the spreading aerial roots.
- Cutting back this evergreen hardy perennial after it flowers will promote new growth.
- Thymes are drought resistant and like well-drained soil in full sun.
- Harvest leaves all year.

Thyme planted near eggplant and cabbages will deter pests. The herb is attractive to bees and produces delicious honey with scents of caraway, lemon, orange, and pine.

Thyme has antiseptic properties and is used as a disinfectant solution for cleaning surfaces. Sachets of thyme may be placed among stored linens and clothes to keep moths at bay.

According to folklore, wild thyme was always blessed by the fairies. Anyone who wished to see the fairies "on a bank where the wild thyme grows" only had to drink an infusion of the herb.

Thyme aids digestion by helping to break down fatty foods. As an ingredient of bouquet garni it is used in soups, stews, and casseroles. It is cooked with chicken, fish, pork, lamb, and roasted root vegetables, and goes into savory dumplings and muffins, and savory jellies. It complements tomato dishes, pasta, and pizza.

Taken as a syrup, thyme benefits respiratory problems such as bronchitis, chest infections, catarrh, coughs, colds, flu, whooping cough, and asthma. A thyme poultice may be used for rheumatism, arthritis, and gout. A thyme salve is used for cuts, bruises, and athlete's foot. A gargle is good for problems of the mouth.

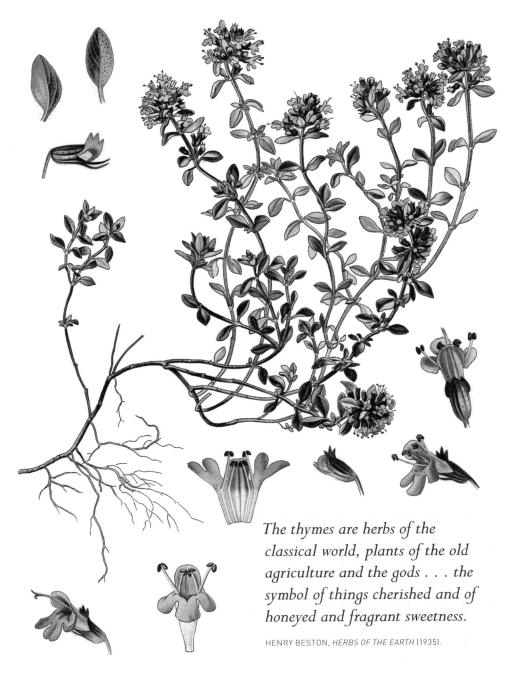

The thymes are herbs of the classical world, plants of the old agriculture and the gods . . . the symbol of things cherished and of honeyed and fragrant sweetness.

HENRY BESTON, *HERBS OF THE EARTH* (1935).

Fenugreek

Fenugreek is native to the Mediterranean and widely naturalized throughout southern Europe, India, Africa, and North America. It has been cultivated for its culinary and medicinal uses and for cattle fodder for thousands of years. The species name, *foenum-graecum*, meaning "Greek hay," was given to it by the Romans; the English common name is derived from the Latin one.

Fenugreek was used by the Egyptians for food and cosmetics, and for inducing childbirth and embalming. In the first century, Dioscorides recommended it for the treatment of a variety of gynecological problems. It was used in Chinese medicine since the eleventh century for its tonic properties, and in Indian Ayurvedic medicine it served as an aphrodisiac and remedy for arthritis and gout. Benedictine monks introduced it to central Europe and grew it in their gardens. The herb was thought to protect from demonic possession.

Fenugreek has long been used to stimulate lactation, and nursing mothers can experience an increase in their milk supply of 900 percent. Modern medical researchers are suggesting that certain components of the plant are potentially advantageous in the treatment of cancer.

- Fenugreek grows easily from seed. Sprinkle the seed in situ once the threat of frost has passed, and cover lightly.
- The plant is a tender annual in temperate climates. It likes full sun and a well-drained, neutral to slightly acid, fertile soil.
- Space plants 6 inches (15 cm) apart.
- The plant will self-seed. Collect seeds when ripe.
- Pick leaves as needed.

Fenugreek plants are rich in natural minerals. They may be used as a "green" manure, although they should be dug in before they flower.

Fenugreek contains iron, phosphorus, and sulfur, and is a good tonic herb for those debilitated by illness and anemia. It soothes sore throats, coughs, and mouth ulcers, and invigorates a dulled sense of taste.

The seeds, both ground and whole, are used in Indian, Moroccan, and Middle Eastern dishes, such as curries and tagines. They are particularly good with potatoes. Fresh leaves are eaten in salads and as a vegetable, and sprouting seeds are also a nutritious addition to salads.

Fenugreek is a skin softener and moisturizer and is used to improve the complexion. Mixed with oil, it is used to massage the scalp and make hair more glossy.

Trillium erectum

Bethroot

Bethroot is an herb native to North America, where it is found growing wild in humus-rich woodlands.

The herb was prized by Native American tribes who used the rhizomes especially to aid childbirth—the common name of bethroot derives from birthroot—as well as menstrual problems and vaginal infections; the old American name of squaw root alluded to the herb's use by women. The Abnaki used it for children, and the Cherokee for gynecological aid, asthma, and cancerous tumors. The Iroquois used the rhizomes and flowers to treat pimples and sunburn.

Early settlers in America, followed by the Shakers, used the herb to treat uterine complaints and reduce excessive menstruation. The root has astringent and antiseptic properties and was used as a lotion by nursing mothers as a remedy for sore, cracked nipples. As a poultice, bethroot was called upon to treat ulcerated sores and gangrene.

Bethroot has one unfortunate characteristic: it has evolved a highly disagreeable odor to attract pollinating flies, like that of putrefied flesh. Folklore has it that just smelling the fresh root is enough to stop a nosebleed. The common name of stinking Benjamin is a reference to the smell.

Picked in spring, the young leaves have the taste of sunflower seeds and are eaten raw in salads, or cooked as a vegetable. The cooked root was taken as an aphrodisiac by Native American tribes. Today, some trilliums are protected by conservation laws.

Bethroot's medicinal uses are mainly historical. It is mentioned by C. S. Rafinesque in his *Medical Flora* (1828–30), and listed in the U.S. National Formulary (1916–47). The herb is astringent, antispasmodic, emetic, and an emmenagogue. It appears in some commercial combination preparations for hemorrhage of the uterus, urinary tract, and lungs, and also for calming excessive menstruation. It is used as a poultice for insect bites, stings, and some snake bites.

Avoid bethroot during pregnancy because it may cause the uterus to contract and result in a miscarriage.

- Bethroot is best propagated by root division in the spring; seed can take up to three years to germinate.
- This hardy, perennial plant prefers humus-rich, slightly acid soil, like that of its woodland habitat.
- Rhizomes can be lifted in the fall.
- The malodorous but pretty flowers bloom from April to June.

Nasturtium

Also called Indian or Mexican cress, the nasturtium is native to Peru and Bolivia and was introduced to Spain by explorers who brought home seeds in the sixteenth century. From there it made its way into Europe and across the Channel to Britain, where John Gerard mentions its use in the 1590s. It was known to him as *Nasturtium indicum*, or "Cresses of India." Having received seeds "from my loving friend John Robin of Paris," he praised it as a "rare and faire plant."

Today, nasturtium is mainly grown for its decorative flowers of bright red, orange, and yellow, and secondarily for its culinary uses. Historically it was valued as a general tonic herb for the digestion, while the seeds were a highly potent purgative for relieving constipation. Nasturtium contains a lot of vitamin C, iron, sulfur, and other minerals, and is thought to possess aphrodisiac qualities, the red-flowered forms especially. Used fresh, the leaves were once used to prevent scurvy.

Called "dry-land watercress," all the plant is edible. The leaves, flowers, and seeds are best eaten raw in salads. The pickled seeds are a good substitute for capers.

An excellent facial wash may be made from infusing the leaves. It benefits stubborn skin problems, mild acne, and pimples. Commercial hair products take advantage of its sulfur content to promote hair growth.

Nasturtiums secrete a strong, pungent essence that repels certain pests in the soil. This invigorates neighboring plants, such as brassicas, pumpkins, and potatoes. Sprayed "nasturtium tea" will control aphids on companion herbs and vegetables.

Antiseptic, antimicrobial, antibiotic, antifungal, and tonic, nasturtium is an excellent healing herb for skin complaints. It is also used for coughs, phlegm, catarrh, and lung problems, and urinary infections.

- Sow seeds in plugs, one in each plug, in early spring.
- Once frosts have passed, plant this annual half-hardy species outside in its final position to grow on through the season.
- It likes a well-drained soil in full sun, in a position where it can sprawl and climb.
- It is widely used for containers and pots, and for edging vegetable gardens and herb patches.
- It self-seeds well, but is definitely not frost hardy.

Tussilago farfara

Coltsfoot

Native to Europe, North Africa, and northern Asia, coltsfoot is naturalized in North America. A rampant traveler, it is found growing on roadsides, verges, hedgerows, and places where the ground has been disturbed, such as quarries and wastelands.

Coltsfoot has a medical history of about 2,500 years, with numerous mentions of its uses as a remedy for coughs, asthma, and bronchial conditions. The genus name is from the Latin *tussis*, meaning a cough; *Tussilago* thus means "cough dispeller." Coltsfoot owes its beneficial effects to its mucilaginous qualities, mainly in the rhizomes. As a tea or syrup, it is anti-inflammatory and a soothing expectorant.

Dioscorides and Pliny both recommended its use as an aid to breathing, and both suggested smoking the leaves to relieve bouts of coughing. This tradition had continued into the modern day, with the herb being included in herbal smoking mixtures for medicinal use.

In the Middle Ages, coltsfoot carried the name of *Filius ante patrem*, "the son before the father," because the flowers appear first, from March onward, and the leaves only when those have faded.

- Coltsfoot can be propagated by seed, but in a herb-garden setting it will invade rampantly by self-seeding and by horizontal spread of the roots.
- This hardy, herbaceous perennial prefers moist conditions. The best way to remove it from a garden is to improve drainage and make the soil too dry for it.
- Coltsfoot is an interesting herb but is best left to grow in wild parts of the garden.

The downy seed heads were once gathered as a soft stuffing for pillows.

The hairs on the plant are irritant. In some countries, coltsfoot is banned or subject to legal restrictions.

Coltfoot leaves may be eaten raw in salads when young, or cooked as a vegetable; they require washing after boiling to remove bitterness. The flowers were once a favorite for making country wine. They were often used with other spring wildflowers in sorbets, rice, soups, fritters, and pancakes.

Anti-inflammatory and expectorant, coltsfoot is used as a decoction, syrup, or infusion for coughs, bronchitis, laryngitis, and respiratory problems, such as emphysema. Externally, it is used for insect bites, skin irritations, ulcers, sores, and varicose veins. The flowers are more potent and effective than the leaves.

Ugni molinae

Chilean Guava

Chilean guava is a member of the myrtle family and has the synonym *Myrtus ugni* Molina. It is native to Chile, where it is known as *murta*, and parts of neighboring southern Argentina. The Valdivian temperate rain forest is its natural habitat. Juan Ignacio Molina, a Chilean Jesuit priest, naturalist, and botanist, gave the herb its species name in 1782.

The aromatic shrub has small, drooping, pale pink, scented flowers that in turn become small, russet-red, applelike fruits. The fruits taste of guava (*Psidium* species) mixed with strawberries, and the plant is also known as the strawberry myrtle. Additionally, the plant is cultivated in Tasmania, where the fruit is marketed as the tazziberry, and in New Zealand, where it is called the ugni berry or New Zealand cranberry.

The Chilean guava was introduced into England in 1844 by botanist William Lobb. The fruit became a favorite of Queen Victoria, and she particularly liked it in the form of "ugni jelly." The fruit's popularity led to its being grown commercially in Cornwall in the Victorian era, but now it is grown only for its ornamental value.

- Grow from cuttings in spring or late summer. The cuttings may be layered.
- The hardy, evergreen, perennial shrub prefers cool conditions, with morning sun and partial shade.
- Chilean guava likes moist, well-drained soil that is slightly on the acidic side.
- It will tolerate temperatures down to 50°F (10°C), but new growth must be protected from late frosts. Bees love the flowers.
- A good plant for woodland and forest gardening.
- The plant is self-fertilizing.

The leaves of Chilean guava may be used as a tea substitute, and the fruits roasted and ground for drinking like coffee. The blueberry-sized fruits are very aromatic and flavorsome, and can be used in cakes, buns, and muffins, or to make jams and jellies. The fruits may be added to gin, as sloes are. They may be mixed with other exotic fruits in a fruit salad, or eaten with cheese as a side fruit. Chilean desserts include *murta con membrillo*, with quince.

For those who remember it [the taste is] very close to Strawberry Space Dust.

MARK DIACONO, *THE TELEGRAPH* WEBSITE

An infusion of the leaves is traditionally used in native herbal medicine for its antioxidant purposes and high vitamin C content.

Urtica dioca ⚠

Stinging Nettle

A wild herb that thrives in most temperate climates of the world, the stinging nettle is particularly fond of damp, nutrient-rich soil. It is common in hedgerows, field edges, grassy areas, and wastelands, where it casts its wind-pollinated seeds far and wide.

Fibers from cloth made from nettles have been found at Bronze Age sites. Even in the twentieth century, the uniforms of soldiers of the German army in World War I were made of spun nettle fibers. In World War II, Britain used a dark green dye obtained from nettles for camouflage, at the same time extracting chlorophyll for use in medicines. Chlorophyll is still extracted for use as a medical and culinary coloring agent.

Northern European countries have long made rope, cloth, and fishing lines from nettle, using the fibers as a substitute for hemp or flax. Dried nettle was once used as highly nutritious fodder for animals and poultry.

In folklore, nettle stings would prevent the influence of sorcery. The ground powder was used to thwart curses, and growing plants near a dairy prevented the milk from being soured by witches.

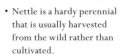

Nettle contains boron and is used for rheumatism. As a diuretic it is used for gout and diabetes. Iron-rich, it is used to treat anemia, and its vitamin C ensures that the iron is well absorbed.

Nettle has lately regained favor as a vitamin- and mineral-rich spring tonic vegetable. It is used for curdling milk in cheese-making, and added to flans, quiches, omelets, soups, beer, pesto, even nettle aloo and nettle spanakopita.

- Nettle is a hardy perennial that is usually harvested from the wild rather than cultivated.
- If required for composting or making liquid feed, nettle should be given its own area.
- Cut off the flowering tops to control prolific self-seeding.
- Harvest only young leaves.

Nettle roots are used as a tonic to promote hair growth and eliminate dandruff. Nettle is added to commercial shampoos and conditioners to impart luster, particularly to dark and gray hair, and is an ingredient of skin-cleansing lotions.

As a dye plant, nettle produces green shades, depending on the season and mordants used. Nettle "tea" is a natural, nitrogen-rich fertilizer that promotes healthy leaf production.

Use only the young, fresh growth for consumption and medicinal purposes.

*Valerian seems to
have a similar effect
to the tranquilizer
benzodiazepine.*

Valerian

Valerian is a native herb of Europe and western Asia, and is naturalized in North America. It prefers stream banks, ditches, and marshy areas but has been found to tolerate drier areas as an escape.

The genus name *Valeriana* may derive from the Latin *valere*, "to be well," although it is also attributed to the Roman emperor Valerius, believed to have been one of the first to use the plant medicinally. It was recommended for use by Hippocrates in the fourth century BCE. It was part of an Anglo-Saxon treatment using leeches in the eleventh century, and throughout the Middle Ages it was used to calm "men that would fight." Nicholas Culpeper recommends the root to be boiled with licorice and aniseed to treat coughs and phlegm, the plague, the stings and bites of venomous creatures, and to expel gas.

The herb helped to soothe those who suffered from the nervous disorder Sydenham's chorea, also known as St. Vitus's dance, and it was traditionally taken as a nerve tonic, sedative, and treatment for insomnia and nervous anxiety. In World Wars I and II, a valerian tincture was administered to troops suffering from shell shock. It is a potent medicine for relaxing the mind and body without becoming addictive.

The secret of the Pied Piper's legendary success in luring vermin from Hamelin is that rats—and cats—are attracted to its smell.

- Sow the small seeds uncovered in plug trays in early spring.
- Plant outside when the roots are sufficiently developed.
- This hardy perennial likes a well-drained, nitrogen-rich soil. The roots need to be kept cool and moist but the plant requires full sun to grow well. Short-term pooling of rainwater is ideal.
- Harvest the rhizome and roots in the fall of the second and third year. These may be dried and stored.
- Valerian makes an attractive, large border plant.

Valerian is particularly soothing when added to bath water for a relaxing soak just before retiring for the night. To make valerian bath oil, add chopped valerian leaves to lavender oil and leave to infuse in a cool place.

Valerian was among the flower species placed in charm bags to attract love (it was also believed to be aphrodisiac) and gain protection, especially against witches. It was also thought to increase psychic ability.

Valerian should not to be taken for extensive periods. As an aid to sleep it should be gradually discontinued after twenty-eight days. Also, it can interact with alcohol, causing overwhelming drowsiness.

Valerian is calmative; it assists in falling asleep and promotes deeper, more restful sleeping. Sedative and nervine, it reduces anxiety and stress. It relieves pain, tension headaches, and vertigo, and also makes a lotion for skin rashes, swollen joints, and varicose veins.

Rich in phosphorus, the leaves are a valuable addition to the compost pile. Earthworms seek out the herb, and thus aerate the soil in its vicinity, a benefit to crops. Sprinkled around and sprayed on sickly plants, a tea made from valerian is said to promote healthy growth.

Vervain

The term "vervain" can be used to refer to the entire *Verbena* genus, which contains about 250 species. The medicinal verbena, *V. officinalis*, also called vervain, is native to Britain, Europe, the Mediterranean, North Africa, and western Asia, and has been widely distributed through many parts of the world, including China, Japan, and North America.

Vervain has been used for many centuries for its medicinal, magical, and mythical properties. The name is derived from the Celtic *fer faen*, which translates as "drive away a stone," a reference to its use in treating stones in the bladder and urethra. It is thought that the Druids held it in high esteem, saying that it opened up the eyes to bring wisdom and inspiration, and magicians used it in love potions and spells to tell the future. *V. officinalis* was also known as Herba veneris, in recognition of its allegedly aphrodisiac properties.

Another name is Herba Sacra, a reference to how it was found growing on Mount Calvary after Christ's crucifixion and assumed to have been used to treat his wounds. Vervain was consequently used in cleansing holy ointments brought to bear against "demonic illness". The herb's sacred use dates from before Christ, however, as it was known as "tears of Isis" to the ancient Egyptians and "Juno's tears" to the Romans.

- Sow seed in spring and plant outside when well rooted.
- Vervain may also be propagated in spring by root division and root cuttings.
- The hardy perennial prefers a well-drained but moist, loamy soil in full sun or part shade.
- Established plants require regular watering or their bloom and growth rates will diminish.
- In receptive conditions, vervain will self-seed prolifically.
- Cut back after flowering.

Nicholas Culpeper regarded vervain primarily as a remedy for women's complaints. It is used today for uterine problems, painful and irregular menstruation, and menopause-related hot flashes, palpitations, and disturbed sleep. Culpeper also used it for gout, yellow jaundice, shortness of breath, and dropsy.

Vervain may cause an upset of the digestive system, or an allergy-related skin rash.

Verveine du Velay, a green liqueur from the Le Puy-en-Velay region of France, is flavored with verbenas and at least ten other herbs, including angelica and sage.

Hot steel used to be hardened by plunging it into cold water. In countries such as Denmark, Finland, and Slovakia, vervain was one of several herbs added to the water to obtain steels of particular qualities. Thus, it was known as "iron herb."

In Chinese medicine vervain was used against malaria. In the West it had uses in alleviating nervous conditions, depression, anxiety, and insomnia. Used externally, it was a treatment for cuts, burns, sores, and wounds. In all, it had a reputation as a cure-all.

At her wedding, a woman would wear a wreath of vervain flowers to preserve her good health and her desire for her husband.

Western European Druids collected verbena when Sirius was rising and the moon was dark.

ENTHEOLOGY WEBSITE

Periwinkle

Periwinkle is native to central Europe and western Asia, and is naturalized throughout North America and parts of the British Isles. It is a woodland plant, and a common name, "joy of the ground," refers to its attractive flowers and its creeping habit. The genus name *Vinca* comes from the Latin *vincire*, "to bind," a reference to its growth habit.

As testified by another common name, "sorcerer's violet," periwinkle has long been associated with magic. It was commonly used in love potions, and was strewn under beds to increase sexual passion. Further, the five-petaled flowers were used as a talisman for protection, and the Anglo-Saxons would hang it in living spaces to ward off evil spirits.

Due to its binding, intertwining character, the French considered periwinkle an emblem of friendship. The Germans construed it as "the flower of immortality," while the Italians called it *fior di morto*, or "the flower of death." In her book, *A Modern Herbal* (1931), the English herbalist Maud Grieve speaks of an ancient Italian custom of making a periwinkle garland to place on the bier of a child who has died.

Despite the plant's toxic nature, Nicholas Culpeper asserts that "the young tops made into a conserve is good for nightmares."

- Propagate by division and layering. The long, trailing stems root at every node as they spread, creating a dense ground cover.
- The hardy perennial prefers loamy soil in semi-shade, set in woodland or hedgerows.
- Remove tall weeds until the periwinkle is established.
- Cut back in fall and winter to prevent over-growth and invasiveness.
- The dense root system makes it a good species for erosion control.

At weddings in western Ukraine, the bride and groom wear crowns of woven periwinkle. With tough, glossy, evergreen leaves that can withstand both freezing temperatures and heat, the plant symbolizes the evergreen nature of their love and marriage vows.

In folk medicine the plant had various uses but was mainly favored for cramps.

If eaten, it causes low blood pressure, nausea, and vomiting.

Periwinkle is astringent and has been used in the past in a mouthwash for gingivitis and mouth ulcers, and in a gargle for sore throats. It increases the flow of blood and oxygen to the brain, and researchers have assessed the vincamine found in the leaves as a potential treatment for arteriosclerosis and dementia. In homeopathy it is used to treat excessive menstruation.

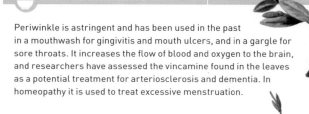

Sweet Violet

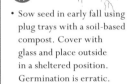

A common wayside plant found in hedgerows and woodlands, sweet violet is native to Asia, North Africa, and Europe, and was introduced to North America and the British Isles.

For many centuries violets have been grown to make perfume, and their seductive, sweet scent symbolizes strong emotions. Violet is a flower of Aphrodite, the goddess of love, and also her son, Priapus, the god of gardens. The Greeks made the violet the symbol of Athens and fertility. The Romans adored violets and added them to their wine, and also wore garlands of violets while drinking to allay the effects of intoxication. Indeed, Horace criticized the Romans for spending too much time growing violets for their wine and neglecting the much more useful olive. Pliny and Hippocrates both believed that violets were good for treating hangovers.

Monks in medieval times grew violet lawns for a place of contemplation. Napoleon was known by his loyal followers as "Caporal Violette" and the flower became the emblem of the Imperial Napoleonic Party. Violets were among Queen Victoria's favorite flowers and were sold on London streets in her reign. On Mothering Sunday, English children were known to "Go a-mothering and find Violets in the lane."

Sweet violet is antiseptic, demulcent, expectorant, and laxative. The flowers are used as an infusion or syrup for insomnia and headache, and to soothe the nerves. Leaf and root infusions are used for catarrh, bronchitis, and hangovers. A violet mouthwash can curb infections. In aromatherapy it is used to treat exhaustion.

Fresh flowers are used as a breath freshener. In a Celtic preparation, violets were steeped in goats' milk to as a beauty enhancer. Violet is widely used for its scent in perfumes and skin lotions, washes, and tonics.

The edible flowers are eaten raw in salads. They are candied for cake decoration, and used to flavor and color confectionery, candy, ice cream, ice cubes, syrups, and creamy dessert puddings.

- Sow seed in early fall using plug trays with a soil-based compost. Cover with glass and place outside in a sheltered position. Germination is erratic.
- Take cuttings with some root attached from the hardy perennial plant in early spring, or plant runners in late spring or early fall. Divide in early summer.
- Violet will self-seed once established.
- Harvest leaves in early spring; flowers as they open; and roots in the fall.

Viola tricolor

Heartsease, Wild Pansy

A flowering wild herb native to Europe and North America, heartsease is also found from western Asia to the Himalayas. It is commonly found in many parts of the world, growing on wastelands, grasslands, and on banks below hedges. The abundant flowers appear throughout the growing season, from spring to fall, and the plant flowers and sets seed all at once, vigorously expelling full seed capsules to perpetuate its existence.

Heartsease is loved, especially by children, for the cheerful, facelike flowers that it generously produces. These droop on their stems for self-protection at night, in heavy dew, and whenever there is any threat of rain. In ancient times, unattached people would grow the plant and use it for love charms and potions, and that is the source of the common name. Old herbals record its dedication to the Blessed Trinity; with three flowers of differing colors, it was known as "herba trinitatis."

Heartsease is anti-inflammatory, diuretic, and expectorant. It was used for disorders of the heart, including a "broken" heart, and for coughs, phlegm, bronchitis, eczema, acne, psoriasis, gout, arthritis, and rheumatism. It reduces fevers and blood pressure.

High doses may cause nausea and vomiting. Do not use heartsease with medications for asthma or with diuretic medicines.

In the garden, heartsease is a desirable companion plant for most vegetables, but especially onions and leeks. Bees are attracted to them, so plant many of them.

The leaves are inedible, but the delicately perfumed fresh flowers may be added to salads, or put into ice pops and cubes for use in cool drinks. The flowers may be crystallized and used to decorate cakes, trifles, and creamy desserts.

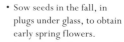

- Sow seeds in the fall, in plugs under glass, to obtain early spring flowers.
- Leave the seeds uncovered to germinate.
- Harden off seedlings before planting outside.
- This hardy perennial is often grown as annual.
- Dead-head the plants to keep the flowers coming.
- Heartsease grows readily in any soil, in full sun or partial shade, and can become rampant if left unchecked.

Wasabi

Wasabi belongs to the pungent Cruciferae/Brassicaceae family, which includes horseradish, radishes, mustards, broccoli, and watercress. *Wasabia* is a genus native to Japan and contains just two species that tend to grow by mountain streams. *W. japonica* has been cultivated since the tenth century in Japan, where is continues to be used medicinally and as a condiment. The cultivar *W. japonica* "Wasabi" is the Japanese equivalent of the West's beloved horseradish (*Armoracia rusticana* syn. *Cochlearia armoracia*), although it can be much more powerful.

Wasabi is now widely cultivated in Japan, and, as a result of increasing international interest in sushi and other Japanese foods, it is also being grown commercially in such countries as England and the United States. In Japan, most of the crop is grown in specific certified areas to ensure its optimal quality, purity, and health. Ironically, the high demand in Japan has necessitated importation of wasabi from producers in China, Taiwan, New Zealand, and the United States.

What appears to be a rhizome root is in fact a buried stem. In Japanese kitchens, wasabi is traditionally prepared with graters—called *oroshigane*—that have a fine grating surface of sharkskin.

- Wasabi is best grown from root division or root cuttings.
- A hardy perennial, wasabi prefers a moisture-retentive, wet soil in semi-shade. It can also grow in water.
- Growing wasabi in a pot is possible, but the plants will die quickly if they are allowed to dry out.
- The use of rain water is recommended.
- Wasabi must be left to grow for three to five years before exploitation as a root crop.

Wasabi roots look like bright green fingers. When grated fresh, the root holds its flavor for a few hours, if covered. Wasabi is eaten with raw fish, in sushi, sashimi, and fish soups, to neutralize any bacteria that may be present in the fish. It is also mixed with soy sauce as a dip. Wasabi can be dried for later use as a paste, or the leaves, stalks, and rhizomes may be preserved by pickling. Commercially prepared wasabi paste is often adulterated with milder ingredients to widen its appeal. The leaves are eaten in spring salads, and add interest to dips.

Wasabi is antibacterial, antifungal, antiviral, and anti-inflammatory. Its use in cuisine originated in its medicinal use in killing food microbes. The sudden rush of vapor it releases when eaten is good for opening the nasal passages.

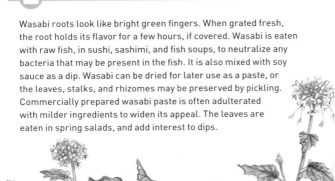

The human tongue is like wasabi: [it] should be used sparingly.

JOHN GREEN,
PAPER TOWNS (2008)

*Eat ginger, and you will love
and be loved as in your youth.*

UNIVERSITY OF SALERNO
MEDICAL SCHOOL, ITALY

Ginger

Ginger is native to Southeast Asia and over many centuries has been introduced to other tropical regions throughout the world. It is now cultivated commercially from Australia to the Caribbean for its thick, fibrous, and strongly aromatic rhizome.

Although ginger is probably best known and used as a spice and flavoring, it is also one of the world's most effective natural medicines. The rhizome compounds change when it is dried, making it twice as pungent and hot, and as medicine fresh and dried ginger are used to treat different ailments. The therapeutic properties of the volatile oils in the rhizome are well researched; gingerol, for example, is known as the constituent responsible for the spice's heat and flavor.

The Romans claimed a tax on the trading of ginger as a commodity. The Chinese have been using ginger medicinally from the time of the Han Dynasty (25–220), and it is used, either fresh or dried, in around 50 percent of their prescribed preparations. In Ayurvedic medicine ginger is known as *vishwabhesaj* or the "universal medicine."

Jamaica has the reputation of producing the best grade of culinary ginger. The Spanish have long been aware of this fact and have been importing Jamaican ginger since the sixteenth century. Jamaica's ginger beer and dark 'n' stormy cocktail with rum are hugely popular.

- Propagate by removing offshoots from the rhizome in late spring.
- Ginger is a tender, perennial, tropical plant. It needs the conditions of a humid tropical forest to reach its full potential.
- Plant in well-drained, humus-rich, neutral to alkaline soil, in a sunny or part-shaded spot.
- As a commercial crop, ginger is treated as an annual or biennial plant.
- A new plant takes a minimum of ten months to grow a rhizome.

Chewed raw or taken in an infusion, ginger can relieve motion sickness, morning sickness, nausea from chemotherapy treatment, and toothache. It stimulates the circulation and promotes perspiration. Ginger is good for breaking a fever, and for colds and flu. It kills parasites and gastrointestinal infections, and relieves some types of food poisoning. It improves liver function and lowers blood pressure. A blood thinner, it helps to prevent heart attacks and strokes.

Ginger should not be taken by anyone with digestive ulcers, a high fever, or an inflammatory skin condition. Exercise caution if pregnant or already taking a blood-thinning medication.

Ginger has an anti-dandruff capability and is used in commercial hair cleansers, shampoos, and conditioners. It is also an ingredient of some facial toners, cleansers, and soaps.

Fresh ginger rhizome is an important ingredient in all Asian cuisines and is used in a multitude of dishes. The skin is always scraped from the rhizome, and the flesh sliced, chopped finely, or grated, depending on the recipe. Young rhizomes are also pickled, and pickled ginger tends to have a pink tinge. Ginger features in many Eastern sweetmeats. In the West it is often crystallized in syrup and candied, often to be eaten at Christmas. It is used in cakes, buns, and cookies.

Glossary

Alkaloids Nitrogen-containing compounds made by plants (but also by other living things), which nearly always have a bitter taste, and sometimes have a potent reaction, and are sometimes toxic. Many of the active ingredients in herbal medical treatments are alkaloids.

Analgesic Pain relieving.

Anti-inflammatory Reduces inflammation.

Antioxidant Prevents or slows down the deterioration of cells by oxidation.

Antispasmodic Reduces spasm or tension, especially in involuntary muscle tissue.

Astringent Causes tissues to contract or shrink and form a protective coating. Astringents also reduce bleeding and mucous discharges.

Calmative Sedative in its effect.

Carminative Relieves flatulence, colic, and other digestive conditions.

Carcinogenic Cancer-inducing. Some chemicals produced by plants may, in susceptible individuals, if ingested over long periods, cause cell damage that is followed by the development of cancers.

Cultivar A variety of a plant where the individuals are genetically identical, and that has been selected for particular desired characteristics, such as large flowers or leaf color. Cultivars are usually propagated by grafting or cuttings.

Decoction A way of extracting dissolved chemicals from herbal or plant material, such as stems, roots, bark, and rhizomes. The plant material is first mashed and then boiled in water to extract oils, volatile organic compounds, and other substances.

Demulcent Soothes and softens tissues that are inflamed, irritated, or damaged, particularly tissues of the digestive tract.

Diaphoretic Causes sweating, which helps to eliminate toxins from the body and reduce fever.

Digestive Having a positive effect on the condition and functions of the digestive tract.

Diuretic Increases the ability of the body to produce urine.

Emetic Causes vomiting.

Emmenagogic Promotes menstruation.

Emollient Softens or soothes the skin.

Essential oil A concentrated liquid containing volatile aroma compounds from plants. Essential oils are also known as volatile oils or simply as the oil of the plant from which they were extracted (such as oil of clove). The oil is called "essential" because it contains the essence of the plant's fragrance.

Expectorant Promotes the expulsion of phlegm from the respiratory tract.

Febrifuge Reduces fever.

Flavonoids Compounds (glycosides) from plants that improve circulation and have diuretic, antispasmodic, and anti-inflammatory effects.

Galactogogue Increases milk flow in nursing mothers.

Genus The first part of the name in the Linnaean system of Binomial Naming (often called scientific or Latin names). A genus collects together closely related species and so may comprise only one or hundreds of species.

Inflorescence A flower head. Popular parlance often refers to such a head of flowers or compound flowers as if they were individual flowers.

Infusion A process of extracting chemical compounds or flavors from a plant material by placing it in a solvent such as water, oil, or alcohol for a period of time (a process often called steeping). The resultant liquid is also called an infusion.

Invasive Having the tendency to spread vigorously, suppressing the growth of other plant species. Usually applied to non-native species, but in some circumstances native species can be invasive in their natural environment.

Laxative Promotes bowel movements.

Macerate Soften or break up a substance using a liquid.

Mucilaginous Secreting a complex, sticky carbohydrate.

Poultice A soft, usually heated substance that is spread on cloth and then placed on the skin to heal a sore or reduce pain.

Purging Strongly laxative in effect.

Relaxant Relaxes tissue that is overactive and tense.

Sedative Reduces anxiety and tension.

Species Plants that are alike and naturally breed with one another, denoted by the second part of the scientific name, after the genus.

Stratification Breaking the dormancy of some kinds of seeds by exposing them to a period of cold.

Subspecies A division of a species, where relatively small differences define a particular geographically defined population as distinct.

Taproot The main downward-growing root of a plant.

Tincture A medicine that is made of a drug mixed with alcohol.

Tisane An infusion (as of dried herbs) used simply as a beverage, or perhaps also for its medicinal effects.

Tonic Improves physiological functions and brings a concomitant sense of well-being.

Umbel A flat-topped or rounded flower cluster in which individual flower stalks grow from a central point on a stem.

Vermifuge Destroys or expels intestinal worms.

Volatile oil See essential oil.

Warming Able to dispel internal cold or hypoactivity and increase vitality, mainly by stimulating the circulation and digestion.

Index

Picture Credits & Acknowledgments

Quintessence would like to thank Cheryl Hunston for the index, Jodie Gaudet for americanizing, and Lesley Malkin for proofreading.

Every effort has been made to credit the copyright holders of the images used in this book. We apologize for any unintentional omissions or errors and would be pleased to insert the appropriate acknowledgment to any companies or individuals in any subsequent editions of the work.

35 De Agostini Picture Library / Contributor 56 Kati Aitken 59 Kati Aitken 105 Florilegius / Contributor 107 Kati Aitken 108 Universal History Archive / Contributor 109 Florilegius / Contributor 110 L.F.J. Hoquart 112 Kati Aitken 124 De Agostini Picture Library / Contributor 130 De Agostini Picture Library / Contributor 135 Kati Aitken 136 DEA / G. CIGOLINI 141 Kati Aitken 159 Kati Aitken 170 © bilwissedition Ltd. & Co. KG / Alamy 175 Dorling Kindersley 215 Valerie Price

Author acknowledgments

Without the encouragement and help that I had from the two most important people in my life, my husband Rob, and my daughter India, who together took on my work at the nursery while I was writing, this book could not have happened. Extra special thanks to little Dot Hall, my mother, who was my secret weapon and has believed in me my whole life. Grateful thanks also goes to Noel Kingsbury for putting my name forward to write this book. I have just one more thank you, which goes to Ralf, my constant four-legged companion.